Laser-based Technologies for Sustainable Manufacturing

This book provides scientific and technological insights on novel techniques of design and manufacturing using laser technologies. It showcases applications of laser micromachining in the biomedical industry, laser-based manufacturing processes in aerospace engineering, and high-precision laser-cutting in the home appliance sector.

Features:

- Each chapter discusses a specific engineering problem and showcases its numerical and experimental solution
- Provides scientific and technological insights on novel routes of design and manufacturing using laser technologies
- Synergizes exploration related to the various properties and functionalities through extensive theoretical and numerical modeling
- Highlights current issues, developments, and constraints in additive manufacturing
- Discusses applications of laser cutting machines in the manufacturing industry and laser micromachining for the biomedical industry

The text discusses optical and laser-based green manufacturing technologies and their application in diverse engineering fields including mechanical, electrical, biomedical, and computer. It further covers sustainability issues in laser-based manufacturing technologies and the development of laser-based ultra-precision manufacturing techniques. The text also discusses the use of artificial intelligence and machine learning in laser-based manufacturing techniques. It will serve as an ideal reference text for senior undergraduate, graduate students, and researchers in fields including mechanical engineering, aerospace engineering, manufacturing engineering, and production engineering.

Advances in Manufacturing, Design and Computational Intelligence Techniques
Series Editor: Ashwani Kumar

The book series editor is inviting edited, reference and text book proposal submission in the book series. The main objective of this book series is to provide researchers a platform to present state of the art innovations, research related to advanced materials applications, cutting edge manufacturing techniques, innovative design and computational intelligence methods used for solving nonlinear problems of engineering. The series includes a comprehensive range of topics and its application in engineering areas such as additive manufacturing, nanomanufacturing, micromachining, biodegradable composites, material synthesis and processing, energy materials, polymers and soft matter, nonlinear dynamics, dynamics of complex systems, MEMS, green and sustainable technologies, vibration control, AI in power station, analog-digital hybrid modulation, advancement in inverter technology, adaptive piezoelectric energy harvesting circuit, contactless energy transfer system, energy efficient motors, bioinformatics, computer aided inspection planning, hybrid electrical vehicle, autonomous vehicle, object identification, machine intelligence, deep learning, control-robotics-automation, knowledge based simulation, biomedical imaging, image processing and visualization. This book series compiled all aspects of manufacturing, design and computational intelligence techniques from fundamental principles to current advanced concepts.

Advanced Materials for Biomedical Applications
edited by Ashwani Kumar, Yatika Gori, Avinash Kumar, Chandan Swaroop Meena, Nitesh Dutt

Additive Manufacturing in Industry 4.0: Methods, Techniques, Modeling, and Nano Aspects
edited by Vipin Kumar Sharma, Ashwani Kumar, Manoj Gupta, Vinod Kumar, Dinesh Kumar Sharma, Subodh Kumar Sharma

Thermal Energy Systems: Design, Computational Techniques and Applications
edited by Ashwani Kumar, Varun Pratap Singh, Chandan Swaroop Meena, Nitesh Dutt

Laser Based Technologies for Sustainable Manufacturing
edited by Avinash Kumar, Ashwani Kumar, Abhishek Kumar

For more information about this series, please visit: https://www.routledge.com/Advances-in-Manufacturing-Design-and-Computational-Intelligence-Techniques/book-series/CRCAIMDCIT

Laser-based Technologies for Sustainable Manufacturing

Edited by
Avinash Kumar
Ashwani Kumar
Abhishek Kumar

CRC Press
Taylor & Francis Group
Boca Raton London New York

CRC Press is an imprint of the
Taylor & Francis Group, an **informa** business

First edition published 2023
by CRC Press
6000 Broken Sound Parkway NW, Suite 300, Boca Raton, FL 33487-2742

and by CRC Press
4 Park Square, Milton Park, Abingdon, Oxon, OX14 4RN

CRC Press is an imprint of Taylor & Francis Group, LLC

ISBN: 978-1-032-39273-8 (hbk)
ISBN: 978-1-032-51470-3 (pbk)
ISBN: 978-1-003-40239-8 (ebk)

DOI: 10.1201/9781003402398

Typeset in Sabon
by MPS Limited, Dehradun

Contents

Aim and Scope   vii

Preface   ix

Acknowledgment   xi

Editors   xiii

Contributors   xvii

1 **Introduction to Optics and Laser-Based Manufacturing Technologies**   1

FRANCIS LUTHER KING M, ASHWANI KUMAR, AVINASH KUMAR, AND ABHISHEK KUMAR

2 **Physics of Laser–Matter Interaction in Laser-Based Manufacturing**   45

ABHISHEK KUMAR, AAYUSH PATHAK, AVINASH KUMAR, AND ASHWANI KUMAR

3 **Current Issues, Developments, and Constraints in Additive Manufacturing**   55

SUSHMA KATTI, V. S. PATIL, G. H. PUJAR, H. J. AMITH YADAV, AND M. N. KALASAD

4 **Laser-Based Additive Manufacturing**   67

PRAMEET VATS, KISHOR KUMAR GAJRANI, AND AVINASH KUMAR

5 **Prospects of AI and ML in Laser-Based Manufacturing Technologies**   85

KHAJA MOHIDDIN SHAIK, YALAVARTHI SURESH BABU, SANJAY GANDHI GUNDABATINI, AND VEDANTHAM RAMACHANDRAN

6 Application of Laser Technology in the Mechanical
and Machine Manufacturing Industry   107

AAYUSH PATHAK, ABHISHEK KUMAR, AVINASH KUMAR,
ASHWANI KUMAR, AND FRANCIS LUTHER KING M

7 Application of Laser-Based Manufacturing Processes
for Aerospace Applications   157

ADITYA PUROHIT, BRIJ MOHAN SHARMA, TAPAS BAJPAI, AND
PANKAJ KUMAR GUPTA

8 Laser Micromachining in Biomedical Industry   169

AVINASH KUMAR, MOHIT BYADWAL, ABHISHEK KUMAR,
ASHWANI KUMAR, AND FRANCIS LUTHER KING M

9 Effect of Laser Surface Melting on Atmospheric
Plasma Sprayed High-Entropy Alloy Coatings   207

HIMANSHU KUMAR, S.G.K. MANIKANDAN, M. KAMARAJ, AND
S. SHIVA

10 Laser Processing Technologies in Electronic and
MEMS Packaging for Advancement of Industry 4.0   235

PAYAL BANSAL, KALPIT JAIN, AND CHETAN DUDHAGARA

Index   249

Aim and Scope

The main objective of the book is to provide researchers and students an intensive study material in the book *Laser-Based Technologies for Sustainable Manufacturing*. The book explains the Lasers applications in advanced materials, cutting-edge manufacturing techniques, innovative design, etc. It covers new evolving trends in engineering related to green technologies, biomedical, computer-aided design, smart manufacturing, AI systems, and sustainability in manufacturing. The present book adopts a balanced approach between academics research and industrial applications.

In the present scenario, manufacturing Industries established a unique place in the community in making the products for home and commercial uses. A lot of manufacturing techniques have been used for producing the engineering products such as milling, grinding, forging, machining, casting, forming, etc., but these techniques are subjected to some factors such as wastage of input material, manpower, and investment. However, improvements in technology over the last few years provided sustainability in most of the manufacturing techniques. In addition, engineers and manufacturing companies also show a strong desire to make products that are capable to provide good service life with an excellent surface finish. In addition, industries that produce these materials face huge pressure to protect the environment during the manufacturing techniques. To address these problems, Laser researchers and other photonics research communities provided novel laser-based technologies for sustainable manufacturing by reducing costs and wastage along with improving the output efficiency and quality of the product. Most importantly, novel and improved laser-based technologies expect to meet community problems and industrial needs. It is most important to show these laser-based technologies must be more flexible, novel, and sustainable.

Laser-based technologies are a non-contact manufacturing process that provides a lot of benefits when compared to the traditional manufacturing process. It has greatly contributed to the modernization of the manufacturing processes. Even though a lot of research efforts are being done to enhance laser-based technologies and explore their applications with sustainable manufacturing. In the future, laser-based technologies need to be automated by reducing men's interface with the system, and the flexibility of the laser-based technologies need to be improved so that it can be easily integrated with other manufacturing processes. Laser-based technologies don't produce any toxic agents. Hence, it is environmentally friendly and safe for all applications.

The book *Laser-Based Technologies for Sustainable Manufacturing* summarizes the past and current work in laser-based non-traditional manufacturing and its application and future scope to green and sustainable manufacturing. This book briefs the pivotal research work of mechanical, electrical, biomedical, and computer science engineering in the area of multidisciplinary research in science and technology. The book has the potential to be valuable to wide readership of researchers, academician, professionals, and graduate students in engineering.

The applicability of this book covers a wide range of industries such as *production sector, design engineer, research and development engineer, automotive industry, aviation sector, electronics industry, nuclear safety research, etc., and it will help to audience conducting research in these industries.* Moreover, each chapter in this book will have literature review, research solution methodology, simulation, experimental setup, and results validation works. The chapters will be well organized and easy to follow. The above help to ensure the completeness of the book and to satisfy the needs of the potential audience in different areas of the world related to manufacturing, design, and green techniques. Having high-quality of research content, it will work as a reference book for researchers and scientists working on solving nonlinear problems in engineering.

The overall goal of this book is to present the latest ongoing research in ten chapters as well as to provide further light on future research work which will be helpful for everyone in the research community.

Dr. Avinash Kumar, Dr. Ashwani Kumar, and Abhishek Kumar

Preface

In 1960, the first laser was made; it was described as *"a solution looking for a problem."* Presently, lasers not only solve problems in production but even provides an entirely novel solution that helps to save energy, material resources, and investment. Here we highlight a few current laser-based technologies for sustainable manufacturing.

In construction projects, stainless steel has been extensively employed materials from building bridges to frames because of its good strength and corrosion resistance in a dry atmosphere. But it greatly gets affected in gaseous and liquid atmospheres. In order to increase the usage of these easily available materials, researchers have employed laser treatments that can improve the strength of stainless steel surfaces. Their team employed laser composite surfacing techniques for improving the properties of stainless steel. These methods not only reduce the wastage of materials and energy consumption but it eliminates the radiation hazards, precisely make the complex shapes and provide defect-free microstructures. This method supports many industries that depend upon stainless steel. Due to the mass production rate in the fashion industries, the requirement for a cost-effective and more efficient process has increased, these requirement has been fulfilled by employing the modern laser technology in the fashion production process. Particularly, cutting of acrylic plastic and metal sheets through conventional processes consumes more time and cost. But while employing the novel laser cutting as a tool for cutting the thicker materials such as acrylic plastic and metals, made it much easier. It also helped to make the unique and complex cuts even on softer materials like thin fabrics without burning the fabric and affecting quality.

In addition, laser engraving techniques have been used to engrave unique designs on thicker materials such as leather and denim. These methods help the footwear and fashion industries not to depend upon unnecessary methods which cause huge stress on the materials to make a precise design. Laser marking is an eco-friendly method used in the automobile industry which helps to mark the products effectively. It helps to avoid the usage of paper, glue, or labels for packing purposes. For example, Companies such as Global retailer Marks and Spencer hope to save ten tonnes of paper and glue every year by using laser marking directly on their fruit and vegetables by making the M&S logo along with product details instead of using paper.

Titanium has been used in many industries such as aerospace, chemical, marine, biomedical, etc., because of its excellent strength, less reactive toward chemicals, corrosion resistance, and biocompatible. But these titanium materials require advanced methodologies to get machined and produce more heat along with more wastage while machining. These factors automatically influence the cost of the material. A team from the University of Johannesburg studied how the titanium alloy responds to laser beam forming. They employed laser beam forming techniques because it is a contact-free method used for precisely shaping the materials, by rapidly heating the localized area on the material's surface with help of a laser beam. It consumes less energy when compared with the conventional heating process. These techniques helped to achieve the desired properties of titanium alloys used in aircraft parts and medical implants by saving the processing time and providing effective control over the heat source power and geometry.

Chapters 1 and 2 detail the basic of laser–matter interaction and laser-based manufacturing technologies. Laser-based additive manufacturing constraint was explained in Chapters 3 and 4. In continuation for sustainable and smart manufacturing use of artificial intelligence and machine learning was explained in Chapter 5.

Chapter 6 explains the uses of laser technology in the mechanical and machine manufacturing industry. In the automobile industry, laser technology has supported to make the laser headlights for cars to replace the traditional tungsten headlights. Tungsten headlights face a lot of disadvantages compared to laser headlights such as tungsten wire getting thinner after a particular period of usage. In addition, it consumes more energy and induces more wastage for manufacturing. Laser headlights are more effective which is four times brighter than traditional LED lights. For example, BMW 7 series uses a series of laser diodes as a power source. Chapters 7, 8, and 10 explain the use of laser technology in aerospace, biomedical, and electronics industries. Chapter 9 explains laser-based surface processing.

The applicability of this book covers a wide range of industries such as *production sector, design engineer, research & development engineer, automotive industry, aviation sector, electronics industry, nuclear safety research, etc., in ten chapters* and it will help to audience conducting research in these industries. Moreover, each chapter in this book will have a literature review, research solution methodology, simulation, experimental setup, and results validation works. It summarizes the past, present, and future work in laser-based sustainable manufacturing.

Avinash Kumar, Ashwani Kumar, and Abhishek Kumar

Acknowledgment

We express our gratitude to CRC Press (Taylor & Francis Group) and the editorial team for their suggestions and support during the completion of this book. We are grateful to all contributors and reviewers for their illuminating views on each book chapter presented in the book *Laser-based Technologies for Sustainable Manufacturing*.

"This book is dedicated to all engineers, researchers, and academicians..."

Editors

Dr. Avinash Kumar is currently working as an Assistant Professor in the Department of Mechanical Engineering at the Indian Institute of Information Technology, Design and Manufacturing (IIITD&M) Kancheepuram, Chennai (An Institute of National Importance under the Ministry of Education, Government of India). He is a Visiting Research Professor at Stanford University California, USA. His research interests include micro/nanofabrication, laser machining, and surface engineering for micro-fluidics and MEMS/Micro/Bio-devices. He has published over 14 international journals, 7 chapters and 9 international conference papers. His google scholar ID is eiJdMdoAAAAJ, ORCID ID is 0000-0002-8071-5748, Scopus ID is 56564412000 and Web of Science Researcher ID is P-6124-2018. He is a reviewer in *ASME Journal of Solar Energy Engineering, Journal of Engineering Applications of Computational Fluid Mechanics, ASME Journal of Heat Transfer, Journal of Physics of Fluids*, and many more. He is a guest editor in "MDPI: Bioengineering" (International Journal) and review editor in "Frontiers in Lab on a Chip Technologies" (International Journal). He worked as a Post-Doctoral Fellow at the Indian Institute of Technology, Kanpur until 2019. His post-PhD research experience includes Early Post-Doctoral Fellow in the Department of Mechanical Engineering, Indian Institute of Technology Delhi. He also worked as an Assistant Professor (TEQIP faculty) in a World Bank and MHRD (Government of India) project (NPIU) for a semester (2018) at THDC-Institute of Hydropower Engineering & Technology, Tehri (Uttarakhand). He obtained his PhD from the Department of Mechanical Engineering, Indian Institute of Technology Delhi in 2018. He received his Bachelor of Engineering in Mechanical Engineering from Rajiv Gandhi Proudyogiki Vishwavidyalaya, Bhopal, in 2010 and an MTech degree in Mechanical Engineering from the Indian Institute of Technology Kanpur, India, in 2012. He was a Research Associate at the Bio-MEMS and Micro-fluidics Laboratory, Indian Institute of Technology Kanpur, India, from August 2012 to November 2012. He worked as Research Associate with Prof. G. K. Ananthsuresh and Prof. Ashitava Ghoshal in the Robert Bosch

Centre for Cyber Physical system, Mechanical Engineering Department, Indian Institute of Science Bangalore for a year (2013). He was qualified for Indian Engineering Services in 2010. He qualified GATE-2010 with 98.20% in 2010. His PhD thesis was nominated for the best thesis award by the foreign and Indian examiners. He was selected for the India–Sri Lanka youth exchange program 2017, organized by the Ministry of youth affairs and sports, Government of India. He was awarded DST-International Travel Fellowship in 2018 to visit Hawaii, USA. He is currently working on various research projects by CSIR and DST-GOI. He has coordinated and organized many conferences and workshops at IIITDM Kanchipuram-Chennai, IIT Delhi, IIT Kanpur, and AIIMS Delhi. He has delivered many talks in Webinars, Faculty Development Programs, and workshops in the institute like (a) B. S. Abdur Rahman Crescent Institute of Science and Technology, Vandalur, Chennai, in association with Tamilnadu State Council for Science and Technology, Chennai, (b) Sathyabama Institute of Science & Technology, Chennai, (c) SRM University Chennai and VIT Chennai. He has examined many MTech and PhD students at SRM University Chennai. He has guided more than ten BTech, MTech, and PhD students for their thesis.

Dr. Ashwani Kumar received PhD (Mechanical Engineering) in the area of Mechanical Vibration and Design. He is currently working as a Senior Lecturer, Mechanical Engineering (Gazetted Officer Group B) at the Technical Education Department, Kanpur, Uttar Pradesh (Under the Government of Uttar Pradesh), India, since December 2013. He worked as an Assistant Professor in the Department of Mechanical Engineering, Graphic Era University Dehradun India from July 2010 to November 2013. He has more than 12 years of research and academic experience in mechanical and materials engineering. He is the series editor of the book series *Advances in Manufacturing, Design and Computational Intelligence Techniques* and *Renewable and Sustainable Energy Developments* published by CRC Press, Taylor & Francis, USA. He is the Editor-in-Chief for the *International Journal of Materials, Manufacturing and Sustainable Technologies (IJMMST)* and the Associate Editor for the *International Journal of Mathematical, Engineering and Management Sciences (IJMEMS)* Indexed in ESCI/Scopus and DOAJ. He is an editorial board member of 4 international journals and acts as a review board member of 20 prestigious (Indexed in SCI/SCIE/Scopus) international journals with high impact factors, i.e., *Applied Acoustics, Measurement, JESTEC, AJSE, SV-JME,* and *LAJSS.* In addition, he has published 100+ research articles in journals, book chapters, and conferences. He has authored/co-authored cum edited 22 books on Mechanical and Materials Engineering. He has published two patents. He is associated with International Conferences as Invited Speaker/Advisory Board/Review Board member/Program Committee Member. He has delivered many invited talks in webinars, FDP, and Workshops. He has been awarded as Best Teacher for excellence in academic and research. He has successfully guided 12 BTech, MTech, and PhD theses. In administration, he is working as a Coordinator for AICTE, E.O.A., Nodal officer for PMKVY-TI Scheme (Government of India), and an Internal Coordinator for CDTP scheme (Government of Uttar Pradesh). He is currently involved in the research area of AI & ML in Mechanical Engineering, Advanced Materials & Manufacturing Techniques, Building Efficiency, Renewable Energy Harvesting, Heavy Vehicle Dynamics and Sustainable Transportation.

Abhishek Kumar is currently working as a PhD Scholar in the Department of Mechanical Engineering, University of California Merced, U.S.A. from January 2021. He received his M.Tech in Design Engineering from Indian Institute of Technology Delhi, India, in 2016. He received his Bachelor of Engineering degree with honors in Mechanical Engineering from Rajiv Gandhi Proudyogiki Vishwavidyalaya, Bhopal, in 2014. He worked as an Assistant Professor in the Department of Mechanical Engineering at Jabalpur Engineering College, Jabalpur, M.P., India, from January 2018 to December 2020. He worked as a summer visiting researcher at the Indian Institute of Technology Kanpur from June 2018 to July 2018. He has previously worked as an Assistant Professor at Galgotias University, Greater Noida (U.P.), from September 2017 to December 2017, and at Parul University, Vadodara (Gujarat), from June 2016 to September 2017. He qualified GATE-2014 with a 99.25 percentile in 2014. His MTech thesis was based on the investigation of nano-wear in Mg-based composites in collaboration with NUS Singapore. He is a Certified Peer Reviewer for Elsevier and has successfully reviewed more than 15 journal papers. He has been recognized as the Outstanding Reviewer by the *Journal of Magnesium and Alloys* (11.90 impact factor) due to his extraordinary contributions to the journal. He is currently involved in the research which involves tribology, mechanical behavior of materials, material characterization, mechanical testing, mechanical properties, microstructure, nanomaterials, material characteristics, materials and materials processing. He has published six research articles in international journals of high impact factors.

Contributors

Yalavarthi Suresh Babu, JKC, Guntur, Andhra Pradesh, India

Tapas Bajpai, Department of Mechanical Engineering, MNIT, Jaipur, Rajasthan, India

Payal Bansal, Poornima College of Engineering, Jaipur, Rajasthan, India

Mohit Byadwal Indian Institute of Information Technology Design and Manufacturing (IIITDM), Kancheepuram, Chennai, India

Chetan Dudhagara, International Agribusiness Management Institute, Anand Agricultural University, Anand, Gujarat, India

Kishor Kumar Gajrani, Indian Institute of Information Technology, Design and Manufacturing, Kancheepuram, Chennai, India; Centre for Smart Manufacturing, Indian Institute of Information Technology, Design and Manufacturing, Kancheepuram, Chennai, India

Sanjay Gandhi Gundabatini, VVIT, Guntur, Andhra Pradesh

Pankaj Kumar Gupta, MNIT, Jaipur Rajasthan India

Kalpit Jain, Poornima College of Engineering, Jaipur Rajasthan India

M. N. Kalasad, Department of Studies in Physics, Davangere University, Davangere, Karnataka, India

M. Kamaraj, Indian Institute of Technology, Madras, India

Sushma Katti, Department of Studies in Physics, Davangere University, Davangere, Karnataka, India; Department of Physics, Research Centre (Recognized by VTU Belagavi), GM Institute of Technology, Davangere, Karnataka, India

Francis Luther King M, SWARNANDHRA College of Engineering and Technology, Narsapur, Andhra Pradesh, India

Abhishek Kumar, University of California, Merced, USA

Avinash Kumar, Indian Institute of Information Technology Design and Manufacturing (IIITDM) Kancheepuram, Chennai, India; Visiting Research Professor, Stanford University California, USA

Ashwani Kumar, Technical Education Department, Uttar Pradesh, Kanpur, India

Himanshu Kumar, Indian Institute of Technology, Jammu & Kashmir, India

S.G.K. Manikandan, Indian Space Research Organization Propulsion Complex, Mahendragiri, India

Aayush Pathak, Indian Institute of Technology Indore, Indore, Madhya Pradesh, India

V. S. Patil, Department of Studies in Physics, Davangere University, Davangere, Karnataka, India

G. H. Pujar, Department of Physics, Research Centre (Recognized by VTU Belagavi), GM Institute of Technology, Davangere, Karnataka, India

Aditya Purohit, MNIT, Jaipur, Rajasthan, India

Vedantham Ramachandran, VVIT, Guntur, Andhra Pradesh

Brij Mohan Sharma, MNIT, Jaipur, Rajasthan, India

Khaja Mohiddin Shaik, VVIT, Guntur, Andhra Pradesh.

S. Shiva, Indian Institute of Technology Jammu, Jammu & Kashmir, India

Prameet Vats, Indian Institute of Information Technology, Design and Manufacturing, Kancheepuram, Chennai, India

H.J. Amith Yadav, Department of Studies in Physics, Davangere University, Davangere, Karnataka, India

Chapter 1

Introduction to Optics and Laser-Based Manufacturing Technologies

Francis Luther King M, Ashwani Kumar, Avinash Kumar, and Abhishek Kumar

CONTENTS

1.1   Introduction 2
1.2   Optics..4
    1.2.1   Concepts of Optics Geometry 5
    1.2.2   Optics Geometry 5
    1.2.3   The Approximation of Ray...6
    1.2.4   Reflection 6
    1.2.5   Law of Reflection...6
    1.2.6   Specular and Diffuse Reflection 7
    1.2.7   Refraction 8
    1.2.8   Law of Refraction 8
1.3   Mirror and Lenses..9
    1.3.1   Mirror..9
        1.3.1.1   Mirror with a Concave Surface....................10
        1.3.1.2   Mirror with a Convex Shape 10
        1.3.1.3   Reflecting in a Parabola 11
    1.3.2   Lens...11
        1.3.2.1   Convex Lens..11
        1.3.2.2   Types of Convex Lenses...............................11
        1.3.2.3   Concave Lens 12
        1.3.2.4   Types of Concave Lens 12
1.4   Laser 12
    1.4.1   Lasers Types...13
        1.4.1.1   Gas-Discharge Lasers 14
        1.4.1.2   Semiconductor Diode Lasers 17
        1.4.1.3   Diode-Pumped Solid State Lasers................17
1.5   Lasers in Manufacturing 18
1.6   Laser Metal Deposition...19
1.7   Lasers in Additive Manufacturing..20
    1.7.1   Use of Lasers in Additive Manufacturing......................20
        1.7.1.1 CO_2 Laser 20
        1.7.1.2   Solid-State Nd: YAG Laser 21

DOI: 10.1201/9781003402398-1

1.7.1.3 Yb-Doped Fiber Laser 22
1.7.1.4 Excimer Gas Laser 23
1.8 Laser Parameters in 3D Printing 23
1.8.1 Operating Wavelength 24
1.8.2 Average Power Pulse Energy Intensity..................... 24
1.8.3 Pulse Duration 25
1.8.4 Beam Quality and Focused Spot Size 25
1.9 Laser for 3D Printing Technology.................................. 26
1.9.1 Stereolithography (SLA) 26
1.9.2 Selective Laser Sintering (SLS)......................... 27
1.9.3 Selective Laser Melting (SLM) 28
1.9.4 Laser-Guided Net Engineering (LENS) 28
1.10 Laser-Based Macro-Scale Metal AM Challenges..................... 29
1.10.1 Interface Defects....................................... 29
1.10.2 Powder Contamination 30
1.10.3 Pre-Processing Software 30
1.10.4 Experimental Design 30
1.11 Industrial Applications 31
1.11.1 Photolithography.. 31
1.11.2 Marking and Scribing 31
1.11.3 Noncontact Measurement 32
1.11.4 Scientific Applications 32
1.11.4.1 Time-Resolved Spectroscopy 33
1.11.4.2 Confocal Scanning Microscopy 33
1.11.4.3 TIR and Fluorescence Correlation
Spectroscopy.................................... 33
1.11.4.4 Microarray Scanning 33
1.11.5 Clinical and Medical Applications 34
1.11.5.1 Flow Cytometry 34
1.11.5.2 Surgical Applications.......................... 34
1.12 Conclusion 35
References 36

1.1 INTRODUCTION

The laser is one of the century's most significant innovations. Numerous scientific, medical, industrial, and commercial uses have been made feasible by the invention of the laser since its debut in 1960. In 1964, Theodore Maiman, the developer of the first functional laser, was reported in *The New York Times* as saying that laser technology was "a solution seeking a problem." These days, it's next to impossible to spend an entire day without touching anything that was made with a laser or that relies on lasers for some aspect of its operation. This Science Resource Guide will delve into the origins of light and optics, the workings of a laser, and the different ways in which lasers have improved our daily lives over the last half-century. Along the way, we'll reflect on pivotal moments in laser history and the pioneering minds that advanced our knowledge of light and the laser. Learn about light and its effects via interactive simulations and hands-on experiments with common household items.

Due to their versatile nature, lasers have ushered in a new era of technological development across the world. The versatility of lasers is shown by the fact that, on the one hand, they may inflict irreparable damage to the eyes, and on the other, they can conduct crucial eye procedures when properly managed. From optical tweezers and laser cooling of atoms in cutting-edge laboratories to laser barcode scanners and laser engravings in everyday life, the full potential of lasers has been apparent over the years. As well as its usage in science and defence, lasers are now increasingly being used in the medical and surgical fields. Lasers are utilised to treat localised sections of tissue and to perform aesthetic surgical procedures in the medical industry. Lasers have found widespread employment in the industrial sector, where they are put to use in a variety of cutting and slicing applications as well as in additive manufacturing processes to enhance the accuracy of 3D printing. The use of lasers as a weapon is so cutting-edge that it's even being used in military operations.

In contrast to regular light, laser light is very powerful, coherent, directed, and monochromatic. Whereas traditional sources show discrepancies in distance, they do not. This allows the full potential of the laser to be transmitted across considerable distances. That's why you can have such high power densities: a lot of energy focused into a little area. It is possible that reflections from lasers might cause irreversible visual loss since people operate them. Traumatic burns, retinal damage, and vaporisation of skin tissue may result from unintentional exposure to high temperatures. Therefore, it is crucial to lessen the amount of laser light that reaches the eye. A guy may look at a 100W tungsten bulb and feel no pain, while exposure to laser light with just 1W of power can cause permanent eyesight loss without warning. There is a rising need for laser safety gear because of the potential dangers posed by expanding laser technology. Based on their output power, lasers are categorised as 2M, 3R, 3B, or 4 kinds. There is a significant need in the realm of cutting-edge research and development for safe laser eyeglasses and associated solutions. Standard laser safety glasses employ one of three materials: polycarbonate filters, absorption glass, or dielectric filters and coatings. Laser safety coatings or filters deposited as a thin layer are one method of defence against laser radiation. Visible light is let through while certain laser wavelengths are blocked by the filter that has been put on the goggles (VLT-visible light transmission). The OD of a filter is its attenuation factor at a given wavelength. When converting from the transmission scale to the OD scale, remember that 100%, 10%, 1%, 0.1%, and 0.0001% all correspond to 0, 1, 2, 3, and 6 on the OD scale.

The visible light threshold (VLT) may be lowered by using absorption glass, whereas polycarbonate filters are effective from low to mid-power-density lasers. Dielectric coatings and filters are superior to these other options because they allow visible light from moderate- to high-power lasers to pass through while simultaneously reflecting back just the unwanted laser wavelengths. According to the laser's output and frequency, these filters may be customised. Due to their direct relationship to eye safety, these filters undergo rigorous quality control examinations. Spectroscopic analysis provides blocking capability of filters in OD and VLT% scales for wavelengths. Industries as varied as aerospace and automotive manufacturing, electronics and semiconductors, medicine, the food and textile industries, building construction, and utility provision all rely on lasers of various types and power levels (CO_2, fibre, solid-state, etc.) for a variety of tasks, including marking, micro- and macro-materials, and other similar tasks. As with many aspects of today's industry, the product's popularity has skyrocketed in recent years. Due to their popularity, laser goggles are seeing a surge in demand, necessitating the use of high-output manufacturing facilities to keep up with supply.

Looking forward, we can see that lasers will play an important role in a wide variety of cutting-edge industries, from sensors and digitalisation to AI and quantum data encryption. To begin, femtosecond laser pulses, which are very brief but extremely powerful, are the wave of the future in medical imaging and other complex procedures that need pinpoint precision. In the production of electric vehicles (EVs), laser welding enables thin and high-speed welding with less heat, both of which are critical in EV batteries. Instead, self-driving vehicles use laser scanning—the laser serving as the car's "eye"—to create a detailed map of the surroundings around it. Additionally, advances in the miniaturisation of semiconductor lasers, diode lasers, and numerous others expand the potential for utilising them in a broad range of complex applications. In conclusion, with the expansion in the invention of laser technology, an awareness of laser safety is important in order to operate with lasers and take benefit.

1.2 OPTICS

Mirrors and lenses may be utilised to bend, divert, and concentrate light, increasing its efficiency. Mirrors and lenses are essential to the functioning of a laser, and they also serve to enhance the device's practical applications. The study of the nature of light and its interactions with materials such as reflection, refraction, and diffraction, falls under the umbrella of optics, a subfield of physics. In addition to lasers, optics is an essential component in the operation of a wide variety of other devices, including cameras, magnifiers, telescopes, microscopes, and many more of the tools we use on a regular basis.

1.2.1 Concepts of Optics Geometry

Whether an optical system is utilised for creating an image or not, geometrical optics is always the starting point. We employ the common sense concepts of a beam of light, which is the route taken by electromagnetic radiation, and surfaces that either absorb or release that radiation. The well-known rule of reflection says that when light is reflected off a smooth surface, the incident and reflected rays create equal angles with the normal to the surface and that the rays and the normal lie in the same plane. As a result of Snell's law, the direction of light rays is altered during transmission. All three directions are coplanar, hence the rule asserts that the sine of the angle between the normal and the incident ray is always equal to the sine of the angle between the normal and the refracted ray.

Concentrators rely heavily on ray tracing, which entails tracking light rays as they travel through a network of reflective and refractive surfaces, for their design and analysis. Although this is a common practice in traditional lens design, the needs of concentrators are unique enough that it will be helpful to declare and build the procedures from scratch. This is because, in a typical lens design, the surfaces that are responsible for reflecting or refracting light are sections of spheres, and the centres of these spheres lie on a single straight line (axisymmetric optical system). This allows for the use of special methods that make the most of the simplicity of the forms of these surfaces and their symmetry. Concentrators used for purposes other than imaging often do not have spherical shapes. Although there is often an axis or plane of symmetry, surfaces may not always have a clear analytical shape. Thus, it will be most practical for us to create shape-specific ray-tracing techniques in which the details are handled automatically by computer programmes based on vector formulations.

1.2.2 Optics Geometry

Light rays are the primary unit of description in geometrical optics. Light rays are conceptualised as geometrical lines starting from sources, extending through media, and being disclosed by detectors; their orientations account for routes along which light travels. The roots of geometrical optics may be found in Maxwell's equations, which are part of the larger theory of electromagnetic. The approximations used in the derivation are shown, and the scope of applicability for geometrical optics is established.

The refractive index n of the medium through which light travels is directly connected to the route that light takes (n is the ratio between the speed of light in a vacuum and in the medium). Assuming n is a smooth continuous function of location, one may write ray equations in general form. More specifically, if n is held constant, rays are just straight lines. Refraction refers to the sudden change in direction of a beam as a consequence of a change in the refractive index at the interface between two materials. Based on such equations, and an appropriate selection of the type, form, and placement of the interfaces, rays may be led to flow in an ordered way via specified media ("lenses"), and finally duplicate the source characteristics ("image"). The fundamental features of optical imaging systems are defined by the first-order approximation of the ray equations. The resultant plan forms the traditional nucleus of geometric optics [1].

1.2.3 The Approximation of Ray

Light is shown to move in a straight path through hundreds of years' worth of observations and tests, up to the point when it comes into contact with the border of another medium. This assertion is often accurate given that the light in question is interacting with things whose wavelengths are much greater than its own. At these length scales, beams of light may be represented as rays, which are either straight lines or lines that have a direction. The ray approximation is a term that describes this kind of simplification. Geometric optics, sometimes known as ray optics, is the subfield of optics that deals with problems that may be approximated using rays. When evaluating the route that light waves take through a variety of materials, we will make use of the ray approximation for the majority of the remaining discussion in this part. When it becomes clear that the ray approximation can no longer be relied upon, we shall move on to discussing the behaviour of waves later in this section.

1.2.4 Reflection

The vast majority of visible objects do not produce their own light but rather reflect light back into the surrounding environment. Light that strikes a surface may be absorbed, transmitted, or reflected. Light is said to be reflected when its wavefront changes direction at the surface of an impermeable or partially permeable substance rather from being absorbed or transmitted completely. Light is absorbed by all substances to some degree, while the remainder is reflected.

1.2.5 Law of Reflection

Let's think about a beam of light that's about to hit a shiny, reflecting surface (Figure 1.1(a)). The angle of incidence is defined as the degree to which an incident ray deviates from a line that is perpendicular to the reflecting surface. The typical describes this particular line. (The term "normal" is commonly used interchangeably with "perpendicular" in physics.) The angle of reflection, or outgoing ray, is r degrees and is measured from the normal. An object's angle of incidence is equal to its angle of reflection, under the law of reflection. This may be expressed symbolically as shown in Figure 1.1(a).

(a)

θ_i θ_r

(b)

Incident rays   Reflected rays

(c)

Incident rays   Reflected rays

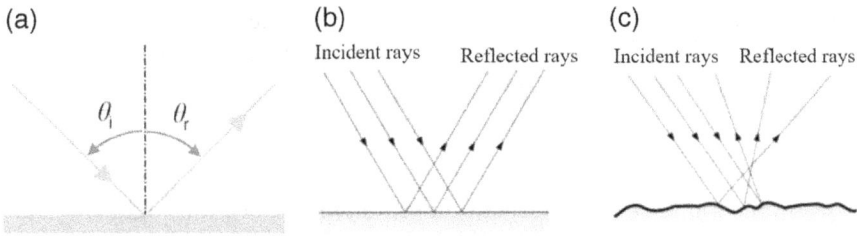

Figure 1.1 (a) Reflecting surface, (b) specular reflection (left), and (c) diffused reflection (right).

Where I is the incident angle and r is the reflected angle, they cancel out to form a single unit. To put it another way, the angle that a reflected light ray makes with the surface is always identical to the angle that the incident light ray makes with the surface.

1.2.6 Specular and Diffuse Reflection

The phenomenon known as specular reflection occurs when light is reflected off of a perfectly smooth surface. In the process of specular reflection, the rays of incident light continue to run in parallel to one another, producing a reflection of the light that is similar to that of a mirror. It goes without saying that a mirror is a good illustration of a specular reflector; nevertheless, specular reflection may also be seen in window glass and calm bodies of water.

It is not necessary for a surface to be perfectly smooth in order for there to be reflection. Diffuse reflection is the term used to describe reflection off of an uneven surface. Because of the uneven surface, each individual light ray is reflected in a little different direction according to the law of reflection, as can be seen in Figures 1.1(b) & (c). This is because the uneven surface causes each individual ray to be reflected in a slightly different direction. The reflection of light off most walls is an example of diffuse reflection; while walls have a tendency to reflect light back into a room, they are not glossy and do not create pictures in the same way as mirrors do. Because of diffuse reflection, we are able to make out the majority of the things in our immediate environment.

The "smoothness" of a surface is related to the wavelength of the wave that is being reflected, and this is how smooth surfaces are measured. If the height of the imperfections on a surface is relatively modest in comparison to the wavelength of the light that is incident on the surface, then the surface will give the appearance of being smooth to the wave, and it will engage in specular reflection. As a consequence of this, it is conceivable for the same surface to give the impression of being rough to one wave while giving the impression of being smooth to another wave with a longer wavelength. For instance, the surfaces of the dishes of huge radio telescopes, some of which have a diameter of more than 50 metres, provide a diffuse reflection of visible light while simultaneously serving as specular reflectors of long-wavelength radio waves.

1.2.7 Refraction

A phenomenon known as refraction occurs when a beam of light travelling through one transparent material meets the border of another transparent material and is refracted in a different direction. Light bends when it encounters a barrier because its speed varies on each side. As light travels through a medium transition, its route is altered, and this change may be explained using Fermat's concept of least time. French mathematician Pierre de Fermat argued in 1662 that light travels the quickest route between two locations. This is the evident straight line joining any two locations made of the same material. If, however, the two materials are vastly different, the light beam will slow down and take a deviated route as it reaches the slower substance. The speed of light in each substance determines the precise direction the beam travels.

1.2.8 Law of Refraction

It is a known fact that in a matter-free vacuum, light always travels at the speed of light, c. The index of refraction describes the amount by which light travel speed is reduced as it travels through a transparent material like air, water, or glass. Calculating a substance's index of refraction, n, is as simple as dividing its speed of light in the material by its speed of light in a vacuum, c:

$$n = \text{speed of light in a vacuum/speed of light in the material} = c/v$$

High refraction indices delay the speed of light. When travelling from a vacuum into a medium having a higher index of refraction, a light beam will experience more bending than when travelling through a substance with a lower index of refraction.

The frequency of light does not alter as it travels through different materials. Without getting too technical, this is because a light wave cannot simply "split" into two halves when it travels through a medium with different refractive indices. In other words, the number of wave crests entering one side of an interface must be equal to the number of wave crests leaving the other side of the interface in a given time interval. Remember from Part I that the relationship between a wave's velocity, frequency, and wavelength is given by the formula $v = f\lambda$. As a result, the wavelength of a light wave must likewise be shrunk if its speed is reduced inside a material that permits it to pass through unimpeded. In a material with index of refraction n, the wavelength λ of a wave with vacuum wavelength will be λ/n.

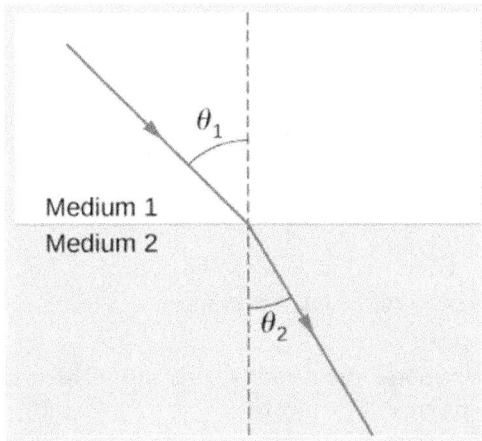

Figure 1.2 Refraction.

W. Snell, a Dutch astronomer, and mathematician from the seventeenth century, is credited with developing the equation that describes the relationship between the angle of incidence and the angle of refraction. Snell's law states that the angle of incidence θ_1 is equal to the refracted angle θ_2 when a beam of light travels from an area with an index of refraction n1 to a region with an index of refraction n2 (Figure 1.2).

1.3 MIRROR AND LENSES

A mirror is an optical device that reflects light rather than letting it flow through it. One side of a glass is silvered to create this material. Plane mirrors and spherical mirrors are both types of mirrors. On the other hand, a lens consists of two separate curves. Allowing light to flow through, while deflecting it. What you're seeing is known as refraction. Both convex and concave lenses serve different purposes.

1.3.1 Mirror

Mirrors are essentially simply a glass surface covered with a metal amalgam on one side, which causes the incoming light beam to reflect rather than bend around the surface. Mirrors, like lenses, may produce pictures with a wide range of characteristics. Face mirrors, for instance, magnify whatever is in front of them. While the reflections in a car's side mirrors or a store's security mirror condense the world into a little image of itself. The angle of incidence is equal to the angle of reflection, as stated by the law of reflection. Virtual pictures of the same size as the item are

(a)

A Parallel rays

At infinity

Principal axis

Reflected rays

C

F

P

B

Focal length

(b)

Reflected rays   Convex mirror

A Parallel rays

Object at Infinity

P

F (Focus)

C

B

Focal length

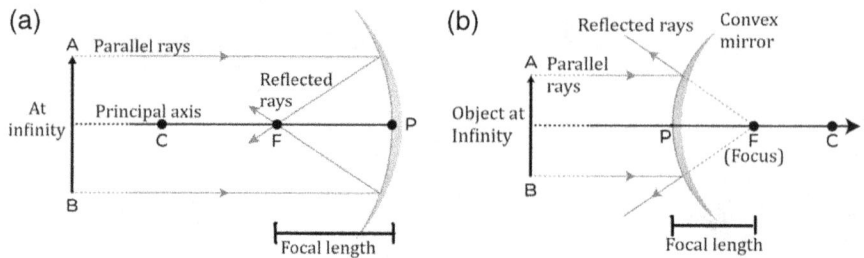

Figure 1.3 (a) Concave surface and (b) convex surface.

always produced by plane mirrors. One distinguishing characteristic is a laterally reversed picture. If you raise your right hand, the mirror image will depict your left hand in that position. Plane mirrors are not only useful in everyday life, but they may also be fashioned into interesting tools like periscopes and kaleidoscopes.

Mirrors may be divided into two distinct categories:

• Concave Mirrors
• Convex Mirrors

1.3.1.1 Mirror with a Concave Surface

Concave mirrors as a special kind of spherical mirror with curved, painted exterior surfaces. The result is a system that can reflect light from inside the surface as shown in Figure 1.3(a). When light travels to a surface and is reflected back, it focuses at a single point. Because of this, the image formed by a concave mirror with a long focal length (the distance of the object must be smaller than the focus length) is enlarged. Makeup mirrors, which need a magnified version of the incident picture, are one practical use of this notion. Dentist mirrors, which magnify the picture of the teeth for better examination, are similarly constructed.

1.3.1.2 Mirror with a Convex Shape

Unlike concave mirrors, which have their outside surfaces painted, convex mirrors have their interior surfaces painted. So, there's a reflective surface on the outside. Light that hits the reflecting surface of this mirror is scattered. Images are created in a virtual world that are upright and shrunken. Convex mirrors are used in the wing mirrors and rearview mirrors of vehicles. Due to the fact that the driver's field of view is increased relative to that of a human's regular vision, this is the case. Allows the driver to keep a closer eye on the road and vehicles approaching from behind.

1.3.1.3 Reflecting in a Parabola

It has the form of the parabola's apex and is, thus, concave. You may use them to focus a beam of light by collecting it off of reflective objects. Since this is the case, it is perfect for use in flashlights and automobile headlights. Parabolic antennas are used for high-energy point-to-point communication, and a tiny bulb is maintained at the focal point of the concave mirror to create parallel beams of light as well.

1.3.2 Lens

An ordinary lens is essentially a flat sheet of transparent material, often glass, with at least one curved surface. It may be used to focus light on a specific location or to diffuse it more broadly.

1.3.2.1 Convex Lens

A convex lens (sometimes called a converging lens) is a lens that is used to focus light rays that are parallel to it and incident on it into a single point behind the lens. The phrase "focal point" describes this precise location. This kind of lens is often seen in eyewear designed to treat farsightedness. Figure 1.4(a) depicts the types of convex lenses.

1.3.2.2 Types of Convex Lenses

1.3.2.2.1 "Plano Convex Lenses"

Because they are only curved on one side, while the other remains flat. Both a curved and a flat surface are present on the positive focal length components of the lens. If you utilise it, you'll get pictures with less spherical distortion. When the subject is distant from the camera's focal point, these lenses come in handy. An infinite conjugate is a concept in optical physics. When you wish to concentrate light from a distant source like a star is a good illustration of this.

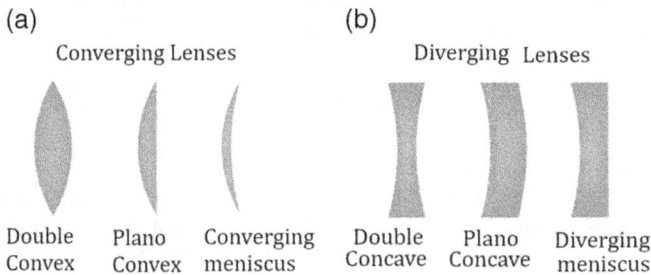

(a) Converging Lenses (b) Diverging Lenses

Double Convex   Plano Convex   Converging meniscus   Double Concave   Plano Concave   Diverging meniscus

Figure 1.4 (a) Convex lenses and (b) concave lenses.

1.3.2.2.2 Convex and Concave Lenses

The outer edges of a double convex lens, sometimes called a biconvex lens or just a convex lens, are curved in the same direction. Light may also be focused with them, however, they work best when used on subjects that are somewhat near to the lens. In optical physics, this is known as a finite conjugate. Microscopy is a common practice in this regard.

1.3.2.2.3 The Concave-Convex Lenses

Often called the Meniscus lens, is a kind of spherical lens with one side curved inward and the other curved outward. It signifies that one side of the lens is concave and the other is convex. The laser beams may be directed using this lens.

Light rays that strike a concave lens are diverted in all directions. These are built with a narrower core and thicker rim. Myopia, or near-sightedness, may be corrected with the use of a convex lens, and these lenses also find usage in torches and spotlights.

1.3.2.3 Concave Lens

When it comes to concave lenses, the ends are often thicker than the rest of the lens. They manipulate the path of incoming light, producing only abstract or reduced-size representations of their subjects. Figure 1.4(b) depicts the types of concave lenses.

1.3.2.4 Types of Concave Lens

1. **A plano concave lens** is a kind of optical lens in which one of its surfaces is flat while the other is concave. With this lens design, the focal length is always less than the distance to the subject. Beam enlargement and light projection are only two of its many applications.
2. **Two-sided concavity**: These lenses are also known as bi-concave lenses because of the fact that their curvature is the same on both the front and back. The focal length of a system may be extended with the use of such lenses [2].

1.4 LASER

Lasers are devices that produce intense, coherent, monochromatic, highly collimated light beams. When compared to other light sources, laser light has an extremely pure (monochromatic) wavelength (colour), and the photons (energy) that comprise the laser beam have a fixed phase relationship (coherence) with respect to one another. Laser light has negligible divergence in most cases. It may focus on a very small area and shine brighter than the sun, or it can travel over vast distances. Because of their properties, lasers are used in a wide variety of applications in all aspects of life.

Charles Townes and Arthur Schalow from Bell Telephone Laboratories came up with the basic ideas for how a laser works in 1958. In 1960, Theodor Maiman at Hughes Research Laboratories showed off the first real laser, which was based on a pink ruby crystal. Since then, thousands of lasers have been made, including one that can be eaten (called the "Jello" laser), but only a small number of them have been used in science, industry, business, or the military. OEMs have found many uses for the helium neon laser (the first continuous-wave laser), the semiconductor diode laser, and the air-cooled ion laser. In OEM applications, the use of diode-pumped solid-state (DPSS) lasers has grown quickly in the past few years.

1.4.1 Lasers Types

(L)ight (A)mplification by (S)timulated (E)mission of (R)adiation is what the word "laser" stands for. To understand what a laser is, you need to know what these words mean. Most people agree that "light" is electromagnetic radiation with a wavelength between 1 nm and 1000 mm. The range of the visible spectrum, which is what we can see, is about 400 to 700 nm. The near-infrared (NIR) is the range of wavelengths from 700 nm to 10 mm, and the far infrared is everything else (FIR). On the other hand, ultraviolet (UV) light is between 200 and 400 nm. Deep UV light is below 200 nm (DUV). Lasers are natural sources of coherent, monochromatic electromagnetic radiation with wavelengths in the UV, visible (VIS), and infrared (IR) ranges (IR). The laser medium could be solid, liquid, or gas, depending on the type of laser and wavelength that was needed.

Depending on whether they send out a continuous wave (cw) or pulses (short or ultrashort) [3–5], they can send out a lot of power or energy. These lasers can be aimed at a precise spot of a certain size on a certain substrate, even if there are other things in the way [6,7]. This means that lasers can be used in many different ways to process materials [8–12].

Maiman [13] made the ruby laser in 1960, for which he won the Nobel Prize. From then until now, laser technology has grown by leaps and bounds. This has led to the development of many new lasers, such as semiconductor lasers, Nd:YAG lasers, CO_2 gas lasers, dye lasers, and the most recent excimer lasers and femtosecond lasers. Materials are removed by Nd:YAG lasers and CO_2 gas lasers through physical processes. Materials are removed by ablation with short pulse lasers like excimer and ultrashort pulse lasers. In the middle of the 1970s, real work on making these industrial lasers started so that different industries could use them to do things like cut, weld, drill, and melt. In the 1980s and early 1990s, lasers were used successfully to heat, clad, alloy, and even deposit thin films in a controlled way. Since then, the technology has grown and changed, and it is now an important part of making parts and objects on a macro, micro, meso, and nanoscale. This has helped many industries, such as semiconductors, traditional manufacturing platforms, the marine, aerospace, and healthcare industries, to name a few.

1.4.1.1 Gas-Discharge Lasers

Gas-discharge lasers are deceptively easy to make: just fill a container with gas, surround it with mirrors, and strike a discharge. In reality, they're far trickier due to the fact that the gas composition, discharge parameters, and container geometry all need to be meticulously planned out to provide the circumstances necessary for a population inversion. Discharge reactions to the container and laser optics must also be carefully considered. The most popular kinds of gas-discharge lasers include noble-gas ion lasers (such as argon and krypton), carbon-dioxide lasers, helium neon lasers, helium cadmium lasers (a metal-vapor laser), carbon-dioxide lasers, excimer lasers, and carbon-dioxide neon lasers. Below, a short discussion will be provided for each of them.

1.4.1.1.1 Helium Neon (HeNe) Lasers

As the second laser ever discovered, the helium neon (HeNe) laser was the first to see widespread commercial usage. Now, only semiconductor diode lasers sell in bigger numbers, although there are already millions of these devices in use.

The HeNe laser uses a low-current (mA) glow discharge at high voltage (kV). While its red (633 nm) output is the most common, HeNe lasers also come in green (543 nm), yellow (594 nm), orange (612 nm), and NIR (1523) varieties. Depending on the wavelength and the size of the laser tube, the output power is just a few tenths to tens of milliwatts.

The gas combination consists mostly of helium (85%), but the neon 15% is what causes the laser to work. Helium atoms are stimulated to a state where they are almost identical to the highest energy levels of neon atoms thanks to the glow discharge. Collisions of the second sort (i.e., stimulating the neon to a higher energy level as opposed to transmitting the energy as kinetic motion) subsequently transfer this energy to the neon atoms. Glow discharges have a negative impedance (i.e., raising the voltage reduces the current), thus a ballast resistor must be connected in series with the laser to make the overall impedance positive so that they may be utilised with a typical current-regulated power supply.

There are five reasons why the HeNe laser is so well-liked and has lasted for so long. they produce a virtually pure single transverse mode beam (M2 1.05), making them the best inherently beam quality laser, many examples of which have an operating life of 50,000 hours or more, they generate relatively little heat and are convection cooled easily in OEM packages, they cost less to acquire and operate, and so on.

1.4.1.1.2 Helium Cadmium (HeCd) laser

A helium cadmium (HeCd) laser is very similar to a HeNe laser, with the distinction that the lasing medium, cadmium metal, is a solid at room temperature. The HeCd laser provides both violet (442 nm) and UV (325 nm) light at low-cost thanks to its cw operation. Its high-quality wavelength match to photopolymer and film sensitivity ranges means it finds widespread usage in holographic and stereolithographic processes working in three dimensions.

In addition to its metallic properties, cadmium also has the property of becoming a solid at room temperature. Evaporating the metal in a reservoir and distributing the vapour evenly down the laser bore are both necessary steps in the lasing process. Electrophoresis is the method used to achieve this goal. Cadmium will plate out on a cold surface, hence great care must be taken in the construction of the laser to keep the cadmium contained and to prevent contamination of the optics and windows, since even a thin coating will add significant losses to halt lasing. When the supply of cadmium is expended, that's generally when things finish.

1.4.1.1.3 Noble-Gas Ion Lasers

High cw power in the visible, UV, and NIR spectrum has traditionally been provided by noble-gas ion lasers (such as argon-ion and krypton-ion). Water-cooled high-power systems are common at universities and research institutes worldwide, whereas air-cooled low-power systems have several OEM uses. Primary output is at 488 nm (blue) and 514 nm (red) for argon-ion lasers, but they may produce up to 7 W in the UV (333–354 nm) and 25 W or more in the visible (454–515 nm) ranges (green). There are three main wavelengths that krypton-ion lasers emit light at 568 nm (yellow), 647 nm (red), and 752 nm (near infrared).

In order to create lasers with a broader spectrum coverage, mixed-gas lasers combine argon and krypton. Ion lasers, in contrast to HeNe lasers, use a high-power, low-pressure arc discharge to do their work (low voltage, high current). A 20-W visible laser will need a power input of 10 kW or more, almost all of which is deposited in the laser head as heat that must be evacuated from the system. The high current densities in the bore (up to 105 A/cm^2) also put a lot of pressure on the materials used in the bore.

Low-power air-cooled lasers and high-power water-cooled lasers (1–20 W) are the two main categories of ion lasers. Both lasers are almost identical in design. Both rely on current being supplied by a coiled, directly heated dispenser cathode, and both have a gas return channel to prevent gas pumping (the uneven buildup of gas pressure along the tube's length, induced by the motion of charged particles toward the electrodes). Bore in air-cooled systems is always built of heat-conducting beryllium oxide (BeO) ceramic. Less than 1 kW of heat is created, and a blower is used to extract it from the outside of the ceramic hole, where it is surrounded by a fin structure. BeO bores and a design in which tungsten discs are fastened to a thin-walled ceramic tube that is encircled by a water jacket are both examples of water-cooled systems that are readily accessible. Water in the vicinity of the discs is heated by conduction from the discs via the tube walls. A solenoid electromagnet completely encircles the bore structure, compressing the discharge to improve current density and reduce bore erosion. When it comes to ion lasers, cathode depletion and gas consumption are the two primary issues that restrict their useful lifespan. The atoms are hurled towards the walls of the discharge tube by the high-velocity discharge and destroyed there. Instability in the discharge is caused by the gradual drop in tube pressure. The use of krypton-ion lasers is complicated by this issue. Systems that use water for cooling often include a way to replenish the water supply to maintain a consistent pressure. However, air-cooled systems often don't, therefore they can only be used for a maximum of around 5000 hours before they become unreliable.

1.4.1.1.4 Carbon dioxide (CO_2) Lasers

Applications where great power is required are where carbon dioxide (CO_2) lasers shine; these lasers can generate substantial amounts of energy with little loss. These lasers typically have a standard output of 10.6 mm and may produce anything from less than 1 W to more than 10 kW of power.

Because CO_2 lasers utilise molecular transitions (vibrational and rotational states) at low enough energy levels that they may be filled thermally, a rise in gas temperature due to the discharge will result in a fall in the inversion level, limiting output power. High-power cw CO_2 lasers mitigate this effect by flushing the discharge zone with cooled (or colder) gas, a process called "flowing gas technology." Since a fresh discharge pulse cannot develop correctly until the hot gas between the electrodes has cooled, this is an especially serious issue with pulsed CO_2 lasers that employ transverse excitation.

CO_2 lasers may come in a wide range of configurations. Fan-assisted transverse gas flow is often used in high-power pulsed and cw lasers to cycle gas through a laminar-flow discharge area, then a cooling region, and finally a lasing region. Small, portable low-power lasers are often made possible by combining waveguide structures with radio-frequency excitation.

1.4.1.1.5 Excimer Lasers

An excimer, also known as an "excited dimer," is a molecular compound of two atoms that is only stable (bound) in an electrically excited state. Only accessible in pulsed form, these lasers provide a very bright output in the deep UV and UV ranges. The XeFl (351 nm), XeCl (308 nm), KrF (248 nm), KrCl (222 nm), ArF (193 nm), and F2 lasers all belong to this group (157 nm). They find widespread usage in fields such as photo-lithography, micromachining, and medicine (refractive eye surgery).

Excimer lasers resemble transverse-flow, pulsed CO_2 lasers in design, at least at first appearance. However, the gases in the system are extremely corrosive, so special care must be taken in the selection and passivation of materials to reduce the corrosive effects of the gases. An infrastructure designed for carbon monoxide would break down within minutes.

The short wavelength of an excimer laser is its main selling point. It is possible to concentrate the excimer beam to a spot size around 40 times smaller than that of a CO_2 laser while maintaining the same quality of light. Furthermore, the excimer lasers with wavelengths approaching 200 nm remove material by ablation (breaking molecules apart) without causing any thermal harm to the surrounding material, while the long CO_2 wave-length destroys material thermally via evaporation (boiling off material).

1.4.1.2 Semiconductor Diode Lasers

Diode lasers directly convert electrical energy into optical radiation by a process known as optical gain, which involves the recombination of injected holes and electrons (and the subsequent emission of photons) in a forward-biased semiconductor pn junction. Diode lasers are capable of converting electrical power to optical power at a rate of up to 50%, which is much higher than that of other lasers by at least an order of magnitude. As the significant hurdles to engineering with diode lasers have been solved over the last 20 years, the trend has been one of a progressive replacement of other laser types by diode laser based-solutions. Diode lasers, due to their small size and low power consumption, have made possible a number of significant new applications, including the realistic implementation of high-speed, broadband transmission of information across optical fibres, a key component of the Internet.

1.4.1.3 Diode-Pumped Solid State Lasers

Diode lasers are efficient, long-lasting, compact light sources, but they present a number of optical challenges that have prompted a new way of thinking about how to make the most of them. These challenges include coupling to the high divergence light, low mode quality in the slow axis of wide-stripe lasers, and low output power from single-transverse-mode lasers. To describe this idea, the phrase "diode-pumped solid-state" (DPSS) laser revolution was coined in the 1980s by a team at Stanford University led by Professor Bob Byer.

There is no complexity to the reasoning. In this setup, a diode laser acts as a pump for an infrared crystal laser, which is then changed to a good mode and its beam is wavelength modulated (using nonlinear optics methods) to produce visible light. Optical pumping of the gain crystal in a conventional high-efficiency infrared laser is done using a discharge lamp, however, this is no longer necessary thanks to the diode laser source. An infrared beam with a good mode is created in the separate resonator, and this beam may be effectively transformed by an intracavity nonlinear crystal into a visible beam with a good mode. A single-mode visible beam is produced with an overall electrical-to-optical conversion efficiency of a few per cent, despite power losses at each stage. These DPSS lasers are now being used in place of the earlier visible gas lasers, whose conversion efficiency seldom exceeds 0.1%.

1.4.1.3.1 Microchip Lasers

One option to get low-power DPSS lasers into mass production without breaking the bank is to model their fabrication after that of semiconductor chips, which would allow for both of these benefits. Using this strategy, MIT Lincoln Labs developed the "microchip" laser in the 1980s. A thin plate of Nd:YVO4 is ground until it is flat, and then it is cut into chips that are 2 mm on a side. Finally, a cube is formed by optically connecting each of these chips to four identical, flat, diced KTP doubling crystal plates. To maximise reflectivity, a 1.06 mm coating is applied to the faces that will be the cube's outer faces before they are diced.

Focusing the light from a single-mode diode laser through the mirrored end of the cube generates enough heat to form a thermally induced waveguide, which in turn forms a stable cavity for infrared lasing. By enclosing the KTP crystal in this cavity, we can transform the IR lasing into a green beam with an output of 10s of milliwatts. In order to keep the pump wavelength and thermal waveguide constant, it is necessary to regulate the diode's temperature. Further, the temperature within the cube has to be maintained at a steady state. The IR laser works at a single longitudinal mode due to the small cavity, and the cavity length has to be thermally regulated to maintain the mode at the maximum of the gain curve. Suppressing "green noise," which will be explained in a moment, is an effect of laser operation at a single frequency.

1.5 LASERS IN MANUFACTURING

Production systems are constantly changing to adapt to the complex demands of the modern era, which is characterised by rapid technological advancements in engineering sectors. These developments are raising the bar for the durability, quality, and performance of production systems and the goods that must function in harsh environments. Industries such as metal forming, aerospace, advanced automotive, and power generation, to name a few, rely heavily on machinery that can withstand high temperatures [14]. Depending upon the uses, the working temperature of a specific produced item may vary up to 500°C and in some circumstances up to 1,000°C. The mechanical properties of the materials and the morphology of their surfaces are both susceptible to damage at these high temperatures (which are often accompanied by thermal fatigue). In turn, this can modify how materials respond to friction, wear, and lubrication [15]. From the perspective of material design, manufacturing, and surface engineering of industrial production systems and products, the harsh operating conditions and complexity of the mechano-chemical processes involved present a great challenge.

Laser-based additive manufacturing (LBAM), according to Ocelk and De Hosson (2018) [16], is a collection of cutting-edge production techniques used to create metal components, fully functioning, and functionally graded products. The manufacturing industry has been changed by LBAM, moving from idea creation to inventive production of application-specific optimised designs [17]. In LBAM, components are made using a powder blown system, laser metal deposition (LMD) system, laser engineered net shaping (LENS) system, powder bed fusion system, selective laser melting (SLM) system, wire-feed system, and wire-laser additive manufacturing (WLAM). The powder-blown systems inject the powder into the laser-induced melt pool using one or more nozzles. A single or several high-intensity laser beams are used in SLM, a powder bed fusion process, to create complex components layer by layer [18]. Instead of employing powder as an energy source, the wire-feed systems (WLAM) create the metal components using a laser beam [19].

1.6 LASER METAL DEPOSITION

Wear plates, also known as guiding plates, are used to keep the rolled material in the proper position throughout the hot strip mill's (HSM) rolling operation. Due to the harsh environment of the HSM, the guiding plates have a short lifespan and need regular maintenance for optimal performance. In an attempt to find a suitable alternative for the reference steel 1.0050, Torres et al. [20,21] studied tribology at high temperatures (up to 700°C). High-temperature sliding wear testing was conducted on the reference material and three substitutes (grey cast iron, abrasion-resistant martensitic grade, and Fe-Cr-V hard-facing material). It was LMD that was used to deposit the hard face material of Fe-Cr-V. The LMD Fe-Cr-V alloy was shown to have the lowest wear rate across all temperature ranges. The lowest wear was the result of the material's ability to retain a high hardness while being heated to higher temperatures.

Lester et al. (2013) [22] compared the tribological performance of alloys laser-generated and traditionally made for continuous caster rolls used in the manufacturing of compact strip using sliding wear tests at ambient temperature and high temperature (at 700°C) (CSP). Low-Moisture-Content (LMD)

WC and high-carbon alloy cast roll steel were tested in room temperature experiments. A higher degree of wear resistance was found in LMD WC compared to that of cast alloy. During tribological testing at temperatures as high as 700°C, LMD Stellite outperformed submerged arc-clad martensitic stainless steel.

1.7 LASERS IN ADDITIVE MANUFACTURING

Additive manufacturing, often known as 3D printing, has emerged as a promising green manufacturing technique in recent years due to its potential for reducing harmful emissions, lowering resource needs, and maximising output. These benefits result from providing energy in discrete bursts, which causes material to be deposited at predetermined target places. The laser beam can transmit a huge amount of energy into a micro-scale focused zone instantly to solidify or cure materials in air, making lasers the most effective energy source in additive manufacturing and allowing for high-precision, high-throughput production of a broad variety of materials.

1.7.1 Use of Lasers in Additive Manufacturing

A laser has three main parts: an optical resonator, a gain medium, and some kind of pumping energy source. The optical resonator's gain medium boosts the light source's output by stimulated emission utilising the pumping source's energy. Solid-state, gas, excimer, dye, fibre, and semiconductor lasers are the most common gain-medium types. These include gas and solid-state lasers, as well as fibre lasers. In light of their prevalence in additive manufacturing and other precision manufacturing applications, CO_2 lasers, Nd:YAG lasers, Yb-fiber lasers, and excimer lasers are discussed here as exemplary lasers.

1.7.1.1 CO₂ Laser

Gas lasers are based on the principle of using a low-pressure gas mixture as the active medium. Electrodes at the tube's beginning and end supply the electric current that excites the gas in a gas laser. The HeNe laser, often known as the HeNe laser, is a popular kind of gas laser that emits light at a wavelength of 632.8 nm, squarely inside the visible spectrum. Lasers that use HeNe cavities are filled with a gas that is a combination of helium and neon. After being energised by an external current, helium atoms collide with neon atoms, bringing them to an excited state that emits light at 632.8 nm. HeNe lasers are ideal for many low-power applications in educational and research facilities due to their affordable price and vivid red output. The CO_2 laser is another kind of gas laser; it's a high-efficiency infrared laser. Lasers fueled by CO_2 are often utilised for high-powered manufacturing tasks like welding and cutting.

An early gas laser, the CO_2 laser was invented in 1964 [23]. Mirrors, windows, and lenses are used to create an optical resonator out of the laser's discharge tube and electric pump source. The gaseous gain medium in CO_2 lasers is CO_2, which is pumped with a direct or alternating current to cause a population inversion and produce laser light [24]. The most popular wavelength for usage in AM, 10.6 m, is produced by CO_2 lasers and has an infrared output of 9.0–11.0 m. Silver or gold is utilised for mirrors, germanium or zinc selenide is used for windows and lenses, and other exotic materials are employed for the optical components due to the infrared wavelength emission [25]. CO_2 lasers are widely utilised in material processing, including but not limited to cutting, drilling, welding, and surface modification, due to their high efficiency (5–20%) and output power (0.1–20 kW) [26]. Between a high reflectivity mirror on one end and a partly reflecting mirror (so-called output coupler) on the other, an electrically pumped gas discharge tube is situated. In addition, for high-power operation of several kilowatts or more, a heat dissipation device like a water jacket to cool the electrodes would be incorporated. CO_2 lasers are the workhorses of precision production because of their cheap cost, excellent dependability, and compactness, all of which result from the system's simplicity. While they can pump a lot of energy into a vast volume of CO_2 gas, their output power isn't very steady since the laser structure expands and contracts as it becomes hot. Gas turbulences in the gas-assisted heat diffusion process may potentially contribute to the instability [27–29]. In high-output use, every two thousand hours, you should inspect the optics for signs of wear and tear [30]. There are a few constraints caused by the long infrared operation wavelength. Because of the CO_2 laser's poor light absorption coefficient in the infrared range, its throughput is restricted to a certain extent when used in the production of metal components. Further, as optical fibres are not capable of providing the wavelength range, CO_2 lasers must rely on free-space bulk reflective optics for beam delivery. Therefore, alternative kinds of lasers must be examined in order to deal with a broader variety of materials or to take advantage of fibre-based beam delivery.

1.7.1.2 Solid-State Nd: YAG Laser

Using rod-shaped Nd: YAG crystals as a solid gain medium, Nd: YAG lasers (neodymium-doped yttrium aluminium garnet laser; Nd3+:Y3Al5O12 laser) are a form of solid-state laser. High-power CO_2 and [31], Nd: YAG lasers are widely employed in a variety of applications. The NIR output wavelength of 1064 nm 28 is achieved by optically pumping the gain medium in the radial direction with a flash lamp or axially with an 808 nm laser diode. One obvious benefit over the CO_2 laser is that the light beam may be supplied by flexible optical fibres at this working wavelength, resulting in a smaller overall system footprint and greater delivery efficiency. Both continuous-wave (when the crystal is lightly doped) and pulsed-mode operations are possible with [29] Nd: YAG lasers (with highly-doped crystals). In continuous mode, the output power is limited to a few kW, but in pulsed mode, it may reach 20 kW at peak power (pulse energy up to 120 J). Due to their reliance on inefficient xenon flash lamps for optical pumping, traditional Nd: YAG lasers have a poor efficiency in converting electrical power into optical radiation. Since most of the unabsorbed energy is lost as heat, the low power efficiency leads to a poor beam quality due to thermal heating of the optical components inducing unanticipated thermal lensing and birefringence effects [32]. The flash lamp's limited lifespan was another area of concern. Diode lasers, also known as diode-pumped solid-state (DPSS) lasers, are an alternative pump source that eliminates these drawbacks [33,34]. The laser's total power efficiency may be increased by a factor of 5 when compared to lamp-pumped lasers due to the improved electrical-to-optical power conversion efficiency of the laser diodes and selective stimulation of the gain medium [35]. As a result of their superior portability and performance, Yb-fiber lasers are quickly replacing Nd: YAG lasers in the additive manufacturing industry. Nd: YAG lasers are still widely utilised in research for parametric study [36–38], or improving production parameters because to their accessibility and widespread availability [39–41]. As an alternative to Nd: YAG lasers, Nd: YVO4 lasers have gained a lot of interest recently due to its larger absorption band, lower operating threshold, and improved efficiency [42,43]. Nd: YVO4 lasers, which have a central wavelength of 1064 nm like Nd: YAG lasers, are largely utilised in SLA [44] as a source of UV light via third harmonic generation (355 nm) to selectively cure photopolymer resins [45,46].

1.7.1.3 Yb-Doped Fiber Laser

A fibre laser is an optical device in which a rare-earth doped optical fibre serves as the active gain medium. In the years after their first creation, fibre lasers were unable to match the output power and pulse energy of bulk lasers. Due to their rapid advancements over the last several decades, however, fibre lasers have emerged as the most viable laser source as an alternative to traditional bulk lasers. Yb-fibers are best for high-power production because of their high quantum efficiency (94%) compared to other rare-earth-doped gain fibres. Because of their superior efficiency, Yb-fiber lasers have largely superseded Nd: YAG lasers in additive manufacturing, where they are utilised for treating a variety of materials [47,48]. Their NIR laser beams have an output wavelength of 1030–1070 nm and are pumped by laser diodes operating between 950 and 980 nm. Because the gain medium and optical components are fibre-based, the system is small, has a high electrical-to-optical efficiency (25%), and is resistant to environmental disturbances. The confines of the fibre itself mean that Yb-fiber lasers are not without their own set of restrictions. Light from bulk lasers travels through air, which is only marginally effective as a light guide. However, as light travels through an optical fibre, the guided light is profoundly impacted by the optical fibre itself, particularly in terms of its nonlinear characteristics. Self-focusing, self-phase modulation, the Kerr lens effect, and the Raman effect are only some of the optical nonlinear processes that might hinder laser performances while working at high peak power [49–52]. Unexpected polarisation shift brought on by fibre bending, vibration, and temperature changes may be just as damaging to laser output [53]. As a gain and light-guiding medium, polarisation-maintaining (PM) optical fibres are suggested for use in more extreme conditions.

1.7.1.4 Excimer Gas Laser

Nanosecond pulses in the UV range are generated by excimer gas laser. Excimer lasers, which employ "excimers" as the gain medium, are driven by pulsed electrical discharge. The term "excimer" is an acronym for "excited dimer," which refers to gas mixes that include a noble gas (such as argon, krypton, or xenon), a halogen (such as fluorine or chlorine), and a buffer gas (typically neon or helium). The most common excimer lasers used in industry produce beams of 193, 248, and 308 nm wavelength, respectively, from a variety of gas mixtures [54]. Like other gas lasers (such as CO_2), an excimer laser requires a pump source, a gain medium, and an optical resonator; in this case, the gain medium is pumped by electrical current. Excimer lasers can only be used in a pulsed mode that generates light bursts with a repetition rate of a few kilohertz and an average output power of a few tens to hundreds of watts. Many optical materials have a high absorptivity around the UV region, making the generation of UV-pulsed light very useful in industrial applications [55–58]. However, excimer lasers are unsuitable because of their low beam quality, complex maintenance, and high operating cost in additive manufacturing equipment [59,60]. Consequently, Nd: YVO4 lasers with their frequencies quadrupled are the best option for generating UV laser beams [61].

1.8 LASER PARAMETERS IN 3D PRINTING

There are several ways in which lasers may be described such as by their average power, power stability, centre wavelength, spectral bandwidth, beam diameter, beam quality, pulse energy, pulse duration, and repetition rate. In order to have a better knowledge of laser systems used in additive manufacturing, it is helpful to categorise the essential factors involved in the process and how they impact production performance. Light-material interaction in most forms of additive manufacturing relies on heat reactions, making laser settings crucial. Here, the sample important parameters are broken down into four categories: wavelength (operating wavelength), power (average power, pulse energy, and intensity), time (pulse duration), and space (pulse size) (beam quality and focused spot size). Due to the interconnected nature of these factors, we shall return to each one individually.

1.8.1 Operating Wavelength

Various materials interact with different laser wavelengths in additive manufacturing. In laser-based additive manufacturing, strong material absorption at the laser wavelength is needed because the target material should interact effectively with the incoming laser light; high absorptivity leads to high production throughput [62–64]. Shorter wavelengths absorb more light in metal powders. Therefore, Nd:YAG or Yb-fiber lasers with a 1064 nm working wavelength have a greater throughput in metal printing. Polymeric materials, a key additive manufacturing material, exhibit substantially greater absorption at 10.6 m than 1064 nm, which explains its broad usage with CO_2 lasers. Operating wavelength affects focusability, which impacts manufacturing resolution. CO_2 lasers are inappropriate for micro/nano-scale production owing to the optical diffraction limit [65].

1.8.2 Average Power Pulse Energy Intensity

Among the many different types of energy sources used to heat up printing materials, lasers are a popular option. Because of this, the laser intensity, which is measured in terms of laser power per unit area, is directly correlated with the output of the process. To begin with, in order to achieve the desired in-situ solidification, sintering, or melting conditions in the target material, the laser intensity must be higher than a predetermined energy threshold [66,67]. While the intensity is linked to the curing or solidification of photopolymer resins, the temperature at which the powder or wire forms is subject to this condition is connected to the sintering or melting point of the material. Some materials, like ceramics, have a very high melting point (zirconium diboride:3245°C), [68] thus they can only be sintered or melted at very high intensities, in contrast to the low melting points of most polymers. Aluminium and copper, which both have high reflectivity and high thermal diffusivity, need even higher intensities to offset the relatively sluggish temperature rise they experience. Once the laser intensity has beyond the production threshold, further increases in intensity may result in a faster construction time. While a stronger laser may raise the build pace, the quality of the manufactured features may suffer if the build rate is too high [69]. It is important to take into account the build pace and the feature quality when selecting the beam power over the threshold energy of the material [70].

When a laser beam is concentrated, its intensity is related to both the average power and the size of the focused spot, both of which are in turn governed by the wavelength at which the laser is operating. For instance, even though CO_2 and Yb-fiber lasers both have the same average power, the concentrated intensity of the Yb-fiber laser may be hundreds of times greater than the CO_2 laser. Due to its shorter wavelength and superior beam quality, a Yb-fiber laser can be concentrated to a significantly smaller area than a CO_2 laser.

1.8.3 Pulse Duration

Depending on where you look in the temporal domain, laser operating modes fall into either a continuous or pulsed category. The output power of a laser operates in one of two modes: either continuously, where it stays constant throughout time, or pulsed, where the output power is released in discrete bursts of a predetermined period. Except for excimer lasers, which can only be used in pulsed mode, all of the lasers considered in this review support both modes. Q-switching, mode-locking, and pulsed pumping all lead to the pulsed mode, which offers much more peak power than the continuous mode. With a pulse length of few ns, an Nd:YAG laser may generate pulses with peak powers in the hundreds of megawatts, enough to melt most target materials in a millisecond of exposure. So, the pulsed mode may provide certain benefits over the continuous mode in laser-based production. High-peak-power light pulses may instantly raise the temperature of the material with little thermal energy loss to the surrounding material, making it simpler to achieve the threshold energy necessary for the machining process. Alternatively, it is more challenging to achieve threshold energies in the cw mode since the same average power would be distributed to the surrounding materials. Additive manufacturing operates on the same basic premise. The goal of SLM is to completely melt the material being irradiated by the laser beam. In order to melt the metal particles, a lengthy pulse duration necessitates a high pulse energy. As a rule, the light-material interaction may be described in terms of heat diffusion, and the threshold of light energy for manufacturing is proportional to the square root of the laser pulse length (from cw to tens of picoseconds) [22].

1.8.4 Beam Quality and Focused Spot Size

The sharpness of the beam and the diameter of the concentrated target area. Laser characteristics in the spatial domain that may be adjusted to enhance manufacturing accuracy include beam quality and concentrated spot size. Beam quality is often defined in additive manufacturing by the "Beam Parameter Product (BPP)," which is the product of the beam radius (measured at the beam waist) and the half-angle of beam divergence (measured in the far field), both expressed in millimetres milliradians (millimetres times milliradians). BPP is strongly associated to power density and impacts manufacturing resolution since low BPP equals high energy confinement. Gain medium, pumping source, resonator geometry, and working wavelength all have a role in this variable. The diffraction-limit, which is a function of the operating wavelength, is the lower limit of BPP. A 1064 nmNd:YAG laser beam has a minimum beam profile profile (BPP) of roughly 0.339 mm rad. The minimal PHP is achieved in theory when the beam profile has a Gaussian distribution; however, because of refractive index gradients, defective optical surfaces, and other perturbing forces, a perfect Gaussian beam does not exist in practice. Beam quality is often defined using the M2 factor (or beam quality factor), which is independent of laser wavelength and so provides a more straightforward method of analysis. The M2 factor equals one if the laser beam is perfectly Gaussian, and is calculated by dividing the BPP by λ/π.

1.9 LASER FOR 3D PRINTING TECHNOLOGY

Ever since lasers were first developed in the 1960s, the manufacturing industry has considered harnessing this kind of tunable optical energy for use in a wide range of operations. Due to their versatility and the increasing demand from consumers, lasers have become an integral part of the industrial sector's fast expansion over the last several decades. Additive manufacturing is one example of a technical breakthrough made feasible by lasers (AM). AM, or 3D printing as it is more often known, has emerged as a major player in the manufacturing sector in recent years. Prototyping and the production of low-volume goods have been profoundly affected by AM [71–78]. It's helped bring computerised CAD (Computer-Aided Design) models and a physically complicated element considerably closer together [79–81]. AM offers a lot of room for creativity in fields where time is money. There has been a lot of interest in laser-assisted manufacturing (LAM) and AM technologies as an alternative to more conventional approaches [74,82–85] because of the lasers' versatility and efficiency. In this part, we will go through SLA, SLS, SLM, and LENS, which are all examples of additive manufacturing procedures that use lasers. According to the American Society for Testing and Materials' (ASTM) "ASTM F42 – Additive Manufacturing" standard, SLA is categorised as a vat photopolymerisation technique, whereas SLS and SLM fall under the category of powder bed fusion, and LENS falls under the category of directed energy deposition. Layer-by-layer fabrication is accomplished using these technologies by employing various lasers and material deposition techniques.

1.9.1 Stereolithography (SLA)

By virtue of its 1984 patenting, SLA may be considered one of the pioneering additive manufacturing techniques [86]. Selective photopolymerisation (SLA) involves shining an UV laser beam over a vat containing photosensitive polymer resin. Before lowering the cured resin layer, the laser beam prints a sharp outline over it.

Among these chemical operations is polymerisation, which is greatly aided by UV light. It is possible for radical or cationic polymerisation to occur in SLA. The photoinitiator of radical polymerisation is a photon absorber that generates free radicals to kick off the polymerisation process. Since photo-initiators have particularly strong absorption in the UV region, the laser source's operating wavelength must be matched to this range for maximum efficiency [87,88]. Anions from acid-containing groups with extremely low nucleophilicity are used in cationic polymerisation. The polymerisation process is kicked off by reactive species formed as a result of UVirradiation. Nd:YVO4 diode-pumped solid-state lasers with a 1064 nm central wavelength are used in commercial SLAsystems, and their wavelength is changed to 355 nm by a third harmonic generation procedure [89]. The critical laser exposure (Ec) for curing photosensitive resins must be greater than a minimum value. The results range from 4.3 to 7.6 mJ/cm^2 for a number of the traditional resins [90]. When using a Gaussian laser source, the cured resin line has the appearance of a parabolic curve. In general, when scanning speed rises or the size of the spot reduces, layer thickness falls as well.

Another method adapted from traditional SLA is micro-stereolithography (SL). Small, complicated items with a micron-scale resolution may be manufactured with SL. Similar to how [91–93] SL works, the photosensitive polymer is "cured" via an external energy source. In most cases, a smaller beam spot size is needed, and the laser energy irradiated on the resin is carefully adjusted such that it is near the critical energy for polymerisation [94]. The employment of neutral absorbers with a highly absorbing reactive medium is feasible. As a result, the polymerised layer may be formed significantly thinner, allowing for improved lateral resolution [95].

1.9.2 Selective Laser Sintering (SLS)

SLS, or selective laser sintering, was created and patented by Carl R. Deckard in the 1980s [96]. Through the solidification of stacked powder layers, additive manufacturing enables the construction of multilayered components and structures. The powder sintering process is assisted by a high-powered laser, which also supplies the necessary heat energy. For layer-by-layer scanning, a laser beam is focused using a beam deflection mechanism (such as a Galvano scanner) to reach the required depth. In SLS operations, CO_2 and Yb-fiber lasers are often employed, while other laser types may be used for certain materials. Polymers have a high absorptivity at the working wavelength of 115–116 nm, hence CO_2 lasers with a few tens to hundreds of Watts of average power are often employed in SLS machines. Direct Metal Laser Sintering (DMLS) is another name for the SLS technique that uses metal rather than polymer as the material of choice [97]. Carbide ceramics may be sintered using a variety of materials, including metal powders, Nd:YAG lasers, and Yb-fiber lasers [98]. The mechanical behaviour and geometry of SLS printed items are also affected by many laser factors in addition to the operating wavelength. The sintering process is mostly affected by the laser power and scanning speed [99]. Target sintering area exposure time is a function of both the laser power and the scanning speed. The processing window for stainless steel-Cu varies depending on whether the laser used is a CO_2 or Nd:YAG. When compared to CO_2 lasers, Nd:YAG lasers have a broader process zone in which laser sintering may occur in stainless steel-Cu. Because more energy is delivered per unit area of the fusion region, the layer thickness of the sintered material tends to grow as the energy density does. Greater sintering causes a greater number of powder particles to fuse together, leading to a thicker layer. An increase in energy density is often accompanied by a corresponding rise in the average density and module strength [100].

1.9.3 Selective Laser Melting (SLM)

Method of SLM, or selective laser melting, is one kind of powder bed fusion technology that involves the use of a laser beam to create three-dimensional objects from metal powder. In a manner similar to SLS, the item is built up layer by layer using laser processing and powder distribution. In SLM, the metal powder is not sintered but rather completely melted as it is laid down in successive layers by a laser of considerably increasing power [101]. Particle bonding is what sets SLM different from SLS [102]. While the entire melting and solidification of powder particles in SLM improves microstructural and mechanical qualities, the process is inherently unstable since the material is constantly changing state from solid to liquid and back again [103]. The melting and solidification process, and hence the qualities of the printed item, is very sensitive to laser parameters including wavelength, repetition rate, pulse duration, and pulse energy. The experiment employed an Nd:YAG laser with power intensities of both 100 W/cm^2 and 250 W/cm^2 with a wavelength of 1.06 m. The powder undergoes dramatic changes in its thermo-physical characteristics at both intensities, leading to an abrupt increase in its absorptivity. The absorptivity was maximised at the thermal equilibrium point after the powders were sintered by surface melting and the particles were rearranged during processing at 100 W/cm^2. While at 250 W/cm^2, the continuous heating causes dramatic melting of particles, followed by the decrease in absorptivity due to the severe loss in porosity. Since metal particles often have better absorptivity at shorter optical wavelengths, Nd:YAG and Yb-fiber lasers, which give shorter wavelength than CO_2 lasers, are favoured in the SLM process. Thin disc lasers and fiber lasers are two examples of high-precision manufacturing lasers due to their superior beam quality [104]. As a result, Yb-fiber lasers have replaced less efficient CO_2 lasers as the light source for most commercial SLM equipment.

1.9.4 Laser-Guided Net Engineering (LENS)

Laser-Guided Net Engineering (LENS) The American Society for Testing and Materials (ASTM) classifies LENS as a kind of additive manufacturing that utilises directed energy deposition. Directed light fabrication, direct metal deposition, laser cast, laser consolidation, laser form, and others are all names for the same technique. Large-scale things that already exist may be maintained and even improved upon by using LENS [104–106]. There is a depth direction (near the focal plane) range of the laser beam with adequate energy density for melting the powders when the laser beam is focussed to a smaller point at the focal plane. This area is referred to as the "critical beam energy-density zone," and is further subdivided into the "hidden spot region" and the "exposed spot region" based on the proximity of the substrate to the focal plane. Deposit height and volume in the melt pool are sensitive to focal plane placement, scanning rate, laser power, and feed rate. The minimum layer thickness of the LENS system should correspond to the thickness of the melt pool. Since LENS systems rely on robotic arms with a large number of degrees of freedom for positioning, Yb-fiber lasers, which allow for very flexible laser beam delivery, are now the most extensively employed kind of laser.

1.10 LASER-BASED MACRO-SCALE METAL AM CHALLENGES

Laser-based macro-scale multiple metallic material AM is still under study. Compared to single-material AM, multi-material AM presents several problems for forming a dense, full, high-performance metallic component. See below.

1.10.1 Interface Defects

Due to the various physical characteristics of several materials, such as melting point, thermal expansion coefficient, and thermal conductivity coefficient, faults are easy to arise during AM. If the thermal properties of the two materials are different, the cooling rate distribution and temperature distribution of treated multi-material components are more difficult. At the interface of different materials, residual thermal stress may produce deformation, delamination, and fissures. If the materials' melting points and laser absorptivities are very different, using the same laser parameters to irradiate dissimilar powder mixture will cause the high melting point powder to be insufficiently melted due to insufficient laser energy density, and the low melting point powder to evaporate due to excessive laser energy density. In the treated material, alloying elements and pores would segregate. Unmelted high-melting-point solids are commonly trapped in the molten pool of low-melting-point metal, causing stress concentration and reducing mechanical performance.

Multi-material additive Manufacturing (MMAM) must also address liquid metal embrittlement. When a solid metal (like stainless steel) encounters a liquid metal (like copper alloy), it becomes brittle. Multiple L-PBF investigations have reported this behaviour. When two metals are melted and combined, hazardous brittle intermetallic compounds may be generated, another MMAM issue. If the density and liquid viscosity of each element in the powder combination are significantly different, the element may not diffuse during AM, resulting to element separation and enrichment zones. During examinations of multiple metallic materials AM, faults at the material interface should be extensively investigated and detected, and preventative steps should be made to minimise or decrease defects.

1.10.2 Powder Contamination

Powder cross-contamination is inevitable in L-PBF because various materials are deposited on the same layer. Material reuse is difficult and needs further study. L-DED requires no additional processes to clean and disseminate powder. Recycling unused powders is difficult. In L-DED, powder splashing is the main cause of contamination, hence laser power and speed must be controlled. Use magnetism [107], particle size distribution [108], material wettability [109], and particle inertia produced by powder density [110] to separate heterogeneous powder mixes. Hybrid wire and powder feeding in L-DED avoids material contamination since the wire is completely melted and unmelted powders may be reused [111].

1.10.3 Pre-Processing Software

These days, STL files are the norm for input in most AM systems. However, commercial pre-processing software for laser-based multi-materials AM is still in its infancy. To accommodate the different material compositions [112] of each region in the placement portion, several experts have developed STL 2.0.

1.10.4 Experimental Design

During AM, flaws such as porosity, delamination, and fractures may easily occur. Defect control is more difficult in an MMAM process. Optimisation of parameters for a single material often requires a significant number of iterative tests. It follows that the time and money spent on tests based on traditional trial-and-error approaches, such as orthogonal experiments and Taguchi experiments, would rise by many orders of magnitude when used to MMAM, particularly AM of FGM. Predicting the ideal process parameters in advance and minimising the actual experimental range requires the development of a multi-material calculating simulation software tool or an artificial intelligence prediction approach based on single-material experimental data. The authors Müller et al. [113] presented a method for identifying novel alloy composition by numerical simulation, which increases the likelihood of successful new alloy creation.

1.11 INDUSTRIAL APPLICATIONS

There is a long history of using high-powered lasers for such applications as material cutting and welding. Lasers are now often employed to engrave numbers and codes on a broad range of objects, as well as to weld the frames of vehicles using robots equipped with lasers. Three-dimensional stereolithography and photolithography are two less common uses.

1.11.1 Photolithography

In the semiconductor industry, lasers play a vital role in exposing photo-resists via masks, which is necessary for the creation of circuits. Before, UV mercury lamps were employed to expose the photoresist, but as feature sizes shrank and more complicated devices were packed onto a single wafer, the mercury lamp's wavelengths became inadequate. Around the turn of the millennium, producers began using UV lasers with approximate wave-lengths of 300 nm to expose the photoresist. In order to achieve the level of detail required by modern semiconductor integrated circuit applications, manufacturers are reducing the wavelengths they use to as little as 193 nm.

1.11.2 Marking and Scribing

Lasers are used to put indelible, human and machine-readable markings and codes on items and packaging. Marking semiconductor wafers for identification and lot control, eliminating the black overlay on numeric display pads, engraving gift goods, and scribing solar cells and wafers are common uses. Laser, scanning head, flat-field focusing lens, and computer control make up the basic marking system. The computer switches on and off the laser beam as it is scanned to form the mark. Depending on the application, scanning may be in a raster pattern (common for dot-matrix markings) or a cursive pattern, with the beam producing letters one at a time. The mark occurs from ablation of the material's surface or photo-chemically induced colour change. Another marking approach, employed with high-energy pulsed CO_2 and excimer lasers, focuses light through a mask holding the marking pattern onto the marking surface.

Laser scribing is similar to laser marking, only the scan pattern is recti-linear and the purpose is to make microscoring along the scan lines. Lasers can mark and scribe metal, wood, glass, silicon, and rubber. Due to crys-talline structure, certain materials exhibit directional absorption and thermal preferences. The kind of laser employed depends on the substance to be marked (e.g., glass transmits 1.06 mm YAG laser output but absorbs 10.6 mm CO_2 laser output). Size, speed, aesthetic quality, and cost are other factors. Most volume marking uses lamp-pumped YAG pulsed or Q-switched lasers. The rest are pulsed and cw CO_2 lasers. DPSS and fibre lasers are approaching due to their improved reliability and reduced cost.

Excimer lasers are utilised in situations demanding high resolution or in which longer wavelengths might harm materials.

1.11.3 Noncontact Measurement

Laser-based noncontact measuring methods include scatter, polarimetry, and interferometry.

a. **Scatter:** Photolithographic methods deposit material patterns on semiconductor wafers. Wafer defects may cause unreliability, circuit disconnects, or circuit failure. Manufacturers must map the wafer to find and eradicate flaws. They use a laser to scan the wafer and a sensitive photodetector array to measure backscatter. Lasers used in this application must have great aiming stability, consistent wavelength and power stability to compute the exact defect size, and low noise so the defect's tiny dispersion can be recognised from the background laser light. Blue 488-nm argon ion lasers are the most popular.

b. **Polarimetry and ellipsometry** can measure the optical phase thickness of a thin layer. A beam with known polarisation and phase arrives at an angle. Thin film refraction index is known. Using the known index of refraction, the phase shift in the reflected beam is connected to the layer's optical phase thickness. This approach may also be employed with thicker transparent media, such as glass, to detect index of refraction fluctuations owing to inclusions or stress-induced birefringence. Violet, red, and NIR single-emitter laser diodes and mid-visible diode-pumped solid-state lasers are often employed because of their cw output, low noise, and small diameters.

c. **Interferometric measurement** may be used for high-resolution location measurement and for measuring optical beam waveform deformation. Using the light beam's wave periodicity as a ruler. The phase of light reflected from an item determines its location in the beam's path. Interference between an object beam and a reference beam produces measurable intensity changes. Moving objects may be measured for distance and velocity by pacing the fringe-recording system. Positioning masks for lithography, mirror distance correlation in an FTIR spectrometer, optical feedback in numerous high-resolution positioning systems, and calculating hard disc drive head alignment and flatness are typical uses.

1.11.4 Scientific Applications

In the scientific lab, lasers play a crucial role in a broad range of spectroscopic and analytical procedures. Confocal scanning microscopy and time-resolved spectroscopy are two such techniques.

1.11.4.1 Time-Resolved Spectroscopy

Time-resolved spectroscopy is a method for studying events that take place in nanoseconds or less. Photosynthesis and other biological processes that take place on time scales on the order of picoseconds (10412 seconds) or less have benefited greatly from this method. The time it takes to see the effect after a fluorescing sample is activated by a laser pulse is substantially less than the pulse's duration. The temporal domain of the fluorescence decay process may then be examined using standard fluorescence spectroscopy measuring methods. Due to the rapid nature of the operations, pulses on the femtosecond (10415 sec) time scale are created using mode-locked lasers as the exciting source, sometimes in conjunction with pulse compression methods.

1.11.4.2 Confocal Scanning Microscopy

When studying a biological specimen, scanning microscopy is utilised to create a three-dimensional picture. A broad volume of the material is lit in conventional light microscopy, and the objective lens collects light not only from the plane in focus but also from above and below it. The final picture incorporates not only the sharpness of the in-focus light but also the blurring effect of the light from the planes that were not in focus. By blocking out the light that isn't directly on the object under study, confocal microscopy is able to create an extremely detailed, high-resolution picture.

1.11.4.3 TIR and Fluorescence Correlation Spectroscopy

The movement of fluorochromes through a predetermined field causes a fluctuation in their fluorescence emission, and this variation may be measured via fluorescence correlation spectroscopy. Once collected, the information may be used to calculate binding and fusion constants for different chemical interactions. Single-photon or two-photon confocal microscopy methods are frequently used due to the tiny volumes being assessed (see above). Fluorescence correlation spectroscopy often focuses on the top 100 to 200 nm of a sample's surface. However, traditional confocal spectroscopy has a 1 to 1.5 mm excitation depth (vertical resolution), resulting in poor signal-to-noise ratios and reduced accuracy.

1.11.4.4 Microarray Scanning

The term "microarray" refers to a matrix of individual DNA molecules connected in sets of known sequence to a substrate that is about the size of a microscope slide and is used in DNA research. Thousands of molecules, each of which is labelled with a unique fluorochrome, may be included in a single array. The array is then read using a microarray reader, where it is probed at each location by a laser with a wavelength in, or very close to, the excitation band of the corresponding protein tags. Measurements of the resultant fluorescence are made, together with records of the positions and sequences of the fluorescent molecules.

1.11.5 Clinical and Medical Applications

Using an argon-ion laser for photocoagulation to close up damaged blood vessels in the retina was one of the first medical uses of lasers. The laser light focuses on the retina, bypassing the lens and vitreous humour, sealing the tear and stopping the haemorrhage. These days, lasers aren't only utilised for vision correction; they're also widely used in analytical instruments, ophthalmology, cellular sorting, and other fields. In the medical field, lasers from throughout the electromagnetic spectrum—from the UV to the infrared—are put to use. This includes not just CO_2 lasers but also solid-state lasers and diode lasers, and an extensive collection of gas lasers.

1.11.5.1 Flow Cytometry

Single-cell quantification is the speciality of flow cytometry. Not only is it a vital instrument in the study of cancer and immunological diseases, but it also finds widespread use in the food business, namely in the testing of natural drinking beverages for the presence of bacteria and other potentially pathogenic microorganisms.

One by one, the cells pass through a capillary or flow cell and onto a laser beam focused on the cell's surface in a basic cytometer. Once inside the cell, the light energy is scattered onto a detector or detector array. The size and form of the cell are in part determined by the distribution and strength of the dispersed energy. Cells are often fluorescently labelled to allow for selective adhesion to cells or cell components with desired properties. Only anybody wearing the tag will glow under the laser. This plays a role in many systems because it facilitates the sorting and separation of cells and other biological components. The most common lasers used in flow cytometry are the blue 488 nm argon-ion laser, the red 632 nm HeNe laser, and the yellow 594 nm HeNe laser. However, new DPSS lasers in a wide range of colours are approaching the market.

1.11.5.2 Surgical Applications

Tissue is sliced, cancers are vaporised, tattoos are removed, plaque and cavities are filled, hair and follicles are obliterated, skin is resurfaced, and eyesight is corrected, to name just a few of the many surgical and dental applications for lasers. Medical applications are quite similar to manufacturing uses. Ablation is used on various materials. In some others, photochemical changes in blood vessels induce shrinking and absorption, while in still others, tissue is sliced or welded to facilitate the desired effect. The key is knowing how different tissues absorb and respond to different wavelengths and intensities of light. With its ability to ablate material from the lens of the eye without inflicting thermal damage that may blur or make the lens opaque, UV excimer lasers are utilised for vision correction. Many tattoo colours are destroyed by radiation around 694 nm, which is why ruby lasers are employed for tattoo removal.

Wrinkles, moles, warts, and discolourations (birthmarks) are often treated using NIR and infrared lasers for cosmetic purposes. In order to facilitate selective absorption at particular areas, photosensitive compounds are typically used topically or injected during these operations. Macular degeneration, caused by the proliferation of blood vessels and scar tissue behind the retina, is another age-related disorder treated with lasers. With this method, the patient receives an injection of a specific dye that improves the blood's ability to absorb laser light. Blood vessels shrink with laser treatment, revealing the retina's metabolic activity. For this purpose, a multi-watt green DPSS laser is often utilised since its green wavelength is not absorbed by the lens or aqueous component of the eye, so limiting its effect to the veins of interest.

Future advancements in laser-based technology will lead to more energy-efficient and sustainable products in all fields of science and technology. This development will be helpful in reductions in cost, power consumption, and size efficiency. Therefore, it is necessary to research the new ideas to get benefits in terms of economy, production and technology [114–120]. In near future, we should expect to see innovative devices using AI, ML, and IoT that improve consumer and patient (biomedical) satisfaction [121–129].

1.12 CONCLUSION

As a result of their excellent performance ratings and low cost, CO_2 and Nd:YAG lasers have been the industrial workhorses for additive manufacturing and other laser-based production processes for quite some time. Nd:YAG lasers are being phased out in favour of Yb-fiber lasers because of their superiority in many respects. These include higher average power, greater system stability, finer parametric tuning, and lower maintenance costs. Excimer lasers may be utilised in additive manufacturing that calls for high-power UV laser beams, despite their worse beam quality and greater cost. Light-matter interaction with varying operating wavelengths, light power, pulse duration, and beam quality was investigated to shed light on the manufacturing performances of laser-based additive manufacturing, including printable material, accuracy, and throughput. Additive manufacturing requires careful consideration when choosing a laser source. Therefore, it is expected that laser-based AM will either continue to displace subtractive manufacturing methods, assist existing methods in becoming more efficient (a process known as "hybrid manufacturing"), or give rise to whole new sectors.

REFERENCES

[1] Molesini, G., *Geometrical Optics, Encyclopedia of Condensed Matter Physics*, Elsevier, 257–267, 2005.

[2] www.bartleby.com/subject/science/physics/concepts/mirrors-and-lenses

[3] Rong, H., Jones, R., Liu, A., Cohen, O., Hak, D., Fang, A. et al., A Continuouswave Raman Silicon Laser. *Nature* 433(7027), 725–728, 2005.

[4] Keller, U., Recent Developments in Compact Ultrafast Lasers. *Nature* 424 (6950), 831–838, 2003.

[5] Emma, P., Akre, R., Arthur, J., Bionta, R., Bostedt, C., Bozek, J. et al., First Lasing and Operation of an Ångstrom-Wavelength Free-Electron Laser. *Nature Photonics* 4(9), 641–647, 2010.

[6] Gattass, R. R. and Mazur, E., Femtosecond Laser Micromachining in Transparent Materials. *Nature photonics* 2(4), 219–225, 2008.

[7] Weck, A., Crawford, T., Wilkinson, D., Haugen, H. and Preston, J., Laser Drilling of High Aspect Ratio Holes in Copper With Femtosecond, Picosecond and Nanosecond Pulses. *Applied Physics A* 90(3), 537–543, 2008.

[8] Schaffer, C. B., Brodeur, A. and Mazur, E., Laser-Induced Breakdown and Damage in Bulk Transparent Materials Induced by Tightly Focused Femtosecond Laser Pulses. *Measurement Science and Technology* 12(11), 1784, 2001.

[9] Zoubir, A., Shah, L., Richardson, K. and Richardson, M., Practical Uses of Femtosecond Laser Micro-Materials Processing. *Applied Physics A* 77(2), 311–315, 2003.

[10] Steen, W., Laser Material Processing—An Overview. *Journal of Optics A:Pure and Applied Optics* 5(4), S3, 2003.

[11] Malinauskas, M., Žukauskas, A., Hasegawa, S., Hayasaki, Y., Mizeikis, V., Buividas, R. et al., Ultrafast Laser Processing of Materials: From Science to Industry. *Light: Science & Applications* 5(8), e16133, 2016.

[12] Perinchery, S., Smits, E., Sridhar, A., Albert, P., van den Brand, J., Mandamparambil, R. et al., Investigation of the Effects of LIFT Printing With a KrF-Excimer Laser on Thermally Sensitive Electrically Conductive Adhesives. *Laser Physics* 24(6), 066101, 2014.

[13] Maiman, T. H., Stimulated Optical Radiation in Ruby. *Nature* 187(4736), 493–494, 1960.

[14] Rahman, M. S., Ding, J., Beheshti, A., Zhang, X. and Polycarpou, A. A., Elevated Temperature Tribology of Ni Alloys Under Helium Environment for Nuclear Reactor Applications. *Tribol. Int.* 123, 372–384, 2018.

[15] Semenov, A. P., Tribology at High Temperatures. *Tribol. Int.* 28, 45–50, 1995.

[16] Ocelík, V. and De Hosson, J. T. M., Thick Metallic Coatings Produced by Coaxial and Side Laser Cladding: Processing and Properties. *Adv. Laser Mater. Process* 2018, 413–459, 2018.

[17] Bandyopadhyay, A. and Traxel, K. D., Invited Review Article: Metal-Additive Manufacturing—Modeling Strategies for Application-Optimized Designs. *Addit. Manuf.* 22, 758–774, 2018.

[18] Parry, L., Ashcroft, I. A. and Wildman, R. D., Understanding the Effect of Laser Scan Strategy on Residual Stress in Selective Laser Melting Through Thermo-Mechanical Simulation. *Addit. Manuf.* 12, 1–15, 2016.

[19] Ding, D., Pan, Z., Cuiuri, D. and Li, H., Wire-Feed Additive Manufacturing of Metal Components: Technologies, Developments and Future Interests. *Int. J. Adv. Manuf. Technol.* 81, 465–481, 2015.

[20] Torres, H., Slawik, S., Gachot, C., Prakash, B. and Rodríguez Ripoll, M., Microstructural Design of Self-Lubricating Laser Claddings for Use in High Temperature Sliding Applications. *Surf. Coatings Technol.* 337, 24–34, 2018a.

[21] Torres, H., Varga, M., Adam, K. and Rodríguez Ripoll, M., The Role of Load on Wear Mechanisms in High Temperature Sliding Contacts. *Wear* 364–365, 73–83, 2016b.

[22] Lester, S., Longfield, N., Griffiths, J., Cocker, J., Staudenmaier, C. and Broadhead, G., New Systems for Laser Cladding. *Laser Tech. J.* 10, 41–43, 2013.

[23] Patel, C. K. N., Continuous-Wave Laser Action on Vibrational-Rotational Transitions of CO2. *Physical Review* 136 (5A), A1187–A1193, 1964.

[24] Witteman, W. J., Continuous Discharge Lasers. *The CO2 Laser*, Springer, 81–126, 2013.

[25] Bass, M., Lasers for Laser Materials Processing. *Laser Materials Processing*, Elsevier, 1–14, 2012.

[26] Witteman, W. J., Introduction. *The CO2 Laser*, Springer, 1–7, 2013.

[27] Tredicce, J. R., Quel, E. J., Ghazzawi, A. M., Green, C., Pernigo, M. A., et al., Spatial and Temporal Instabilities in a CO2 Laser. *Physical Review Letters* 62 (11), 1274–1277, 1989.

[28] Nighan, W. L., Wiegand, W. J. and Haas, R. A., IonizationInstability in CO2 Laser Discharges. *Applied Physics Letters* 22 (11), 579–582, 1973.32. Digonnet, M., Gaeta, C. and Shaw, H., "1.064- and 1.

[29] umNd:YAG Single Crystal Fiber Lasers. *Journal of Lightwave Technology* 4 (4), 454–460, 1986.

[30] Farças, I. I., "Development of Laser Material Processing inRomania", Springer, pp. 283–290, 1993.

[31] Geusic, J. E., Marcos, H. M. and Van Uitert, L., Laser Oscillationsin Nd-doped Yttrium Aluminum, Yttrium Gallium and GadoliniumGarnets. *Applied Physics Letters* 4 (10), 182–184, 1964.

[32] Weber, R., Neuenschwander, B. and Weber, H. P., Thermal Effectsin Solid-state Laser Materials. *Optical Materials* 11 (2), 245–254, 1999.

[33] Berger, J., Hoffman, N. J., Smith, J. J., Welch, D. F., Streifer, W., et al., Fiber-bundle Coupled, Diode End-pumped Nd:YAG Laser. *Optics Letters* 13 (4), 306–308, 1988

[34] Zhou, B., Kane, T. J., Dixon, G. J. and Byer, R. L., Efficient,Frequency-stable Laser-diode-pumped Nd:YAG Laser. *OpticsLetters* 10 (2), 62–64, 1985.

[35] Hügel, H., New Solid-state Lasers and Their Application Potentials. *Optics and Lasers in Engineering* 34 (4), 213–229, 2000.

[36] Kruth, J. P., Kumar, S. and Van Vaerenbergh, J., Study of Laser-Sinterability of Ferro-based Powders. *Rapid Prototyping Journal* 11 (5), 287–292, 2005.

[37] Mumtaz, K. and Hopkinson, N., Selective Laser Melting ofInconel 625 Using Pulse Shaping. *Rapid PrototypingJournal* 16 (4), 248–257, 2010.

[38] Kobryn, P. A. and Semiatin, S. L., The Laser AdditiveManufacture of Ti-6Al-4V. *JOM* 53 (9), 40–42, 2001.

[39] Liao, H. and Shie, J., Optimization on Selective Laser Sinteringof Metallic Powder via Design of Experiments Method. *RapidPrototyping Journal* 13 (3), 156–162, 2007.

[40] Balla, V. K., Bose, S. and Bandyopadhyay, A., Processing of BulkAlumina Ceramics Using Laser Engineered Net Shaping. *International Journal of Applied Ceramic Technology* 5 (3), 234–242, 2008.

[41] Garg, A., Lam, J. S. L. and Savalani, M. M., Laser Power Based Surface Characteristics Models for 3-D Printing Process. *Journal of Intelligent Manufacturing*, 1–12, 2015.

[42] Minassian, A., Thompson, B. and Damzen, M. J., Ultrahigh-Efficiency TEM00 Diode-side-pumped Nd:YVO4 Laser. *Applied Physics B* 76 (4), 341–343, 2003.

[43] Fields, R. A., Birnbaum, M. and Fincher, C. L., Highly EfficientNd:YVO4 Diode-laser End-pumped Laser. *Applied Physics Letters* 51 (23), 1885–1886, 1987.

[44] Humphreys, H. and Wimpenny, D., Comparison of Laser-basedRapid Prototyping Techniques. Seventh International Conferenceon Laser and Laser Information Technologies, 407–413, 2002.

[45] Huang, B. W., Weng, Z. X. and Sun, W., Study on the Propertiesof DSM SOMOS 11120 Type Photosensitive Resin for Stereolithography Materials. *Advanced Materials Research*, 194–197, 2011.

[46] Huang, B. W. and Chen, M. Y., Evaluation on Some Properties ofSL7560 Type Photosensitive Resin and its Fabricated Parts. *Applied Mechanics and Materials* 117, 1164–1167, 2012.

[47] Orlan Hayes, Marking with Fiber Lasers. *Industrial Laser solutions* 27 (5), 2012.

[48] Verhaeghe, G. and Hilton, P., Battle of the Sources—Using a High-power Yb-fibre Laser for Welding Steel and Aluminium. *Proceedings of the IIW Conference on Benefits of New Method and Trends in Welding to Economy, Productivity and Quality* 10, 15, 2005.

[49] Gu, G., Kong, F., Hawkins, T., Parsons, J., Jones, M., et al., Ytterbium-doped Large-mode-area All-solid Photonic Bandgap Fiber Lasers. *Optics Express* 22 (11), pp. 13962–13968, 2014.

[50] Gu, G., Kong, F., Hawkins, T. W., Foy, P., Wei, K., et al., Impactof Fiber Outer Boundaries on Leaky Mode Losses in Leakage Channel Fibers. *Opt. Express* 21 (20), 24039–24048, 2013.

[51] Kong, F., Gu, G., Hawkins, T. W., Parsons, J., Jones, M., et al., Flat-top Mode from a 50 μm-core Yb-doped Leakage ChannelFiber. *Optics Express* 21 (26), 32371, 2013.

[52] Limpert, J., Schreiber, T., Nolte, S., Zellmer, H., Tünnermann, A., et al., High-power Air-clad Large-mode-area Photonic Crystal Fiber Laser. *Optics Express* 11 (7), 818–823, 2003.

[53] Sezerman, O. and Best, G., Accurate Alignment Preserves Polarization. *Laser Focus World* 33 (12), S27–S30, 1997.

[54] Basting, D., Pippert, K. D. and Stamm, U., History and Future Prospects of Excimer Lasers. Second International Symposiumon Laser Precision Micromachining. International Society forOptics and Photonics, 25, 2002.

[55] Mann, K. R. and Eva, E., Characterizing the Absorption and Aging Behavior of DUV Optical Material by High-resolution Excimer Laser Calorimetry. 23rd Annual InternationalSymposium on Microlithography, 1055–1061, 1998.

[56] Jaber, H., Binder, A. and Ashkenasi, D., High-efficiency Microstructuring of VUV Window Materials by Laser-induced Plasma-assisted Ablation (LIPAA) with a KrF Excimer Laser. *Lasers and Applications in Science and Engineering*, 557–567, 2004.

[57] Morozov, N. V., Laser-induced Damage in Optical Materialsunder UV Excimer laser radiation. *Laser-Induced Damage in Optical Materials: 1994. International Society for Optics and Photonics*, 153–169, 1995.

[58] Wang, X., Shao, J., Li, H., Nie, J. and Fang, X., Analysis of Damage Threshold of K9 Glass Irradiated by 248-nm KrF ExcimerLaser. *Optical Engineering* 55 (2), 27102-27102, 2016.

[59] Lee, K. and Lee, C., Comparison of ITO Ablation Characteristics Using KrF Excimer Laser and Nd:YAG Laser. SecondInternational Symposium on Laser Precision Micromachining.International Society for Optics and Photonics, 260–263, 2002.

[60] Atezhev, V. V., Vartapetov, S. K., Zhukov, A. N., Kurzanov, M. A. and Obidin, A. Z., Excimer Laser with Highly Coherent Radiation. *Quantum Electronics* 33 (8), 689–694, 2003.

[61] Toenshoff, H. K., Ostendorf, A., Koerber, K. and Meyer, K., Comparison of Machining Strategies for Ceramics Using Frequency-converted Nd:YAG and Excimer Lasers. Second International Symposium on Laser Precision Micromachining. International Society for Optics and Photonics, 408–411, 2002.

[62] Gu, D. D., Meiners, W., Wissenbach, K. and Poprawe, R., Laser Additive Manufacturing of Metallic Components: Materials, Processes and Mechanisms. *International Materials Reviews* 57 (3), 133–164, 2012.

[63] Garban-Labaune, C., Fabre, E., Max, C. E., Fabbro, R., Amiranoff, F., et al., Effect of Laser Wavelength and Pulse Duration on Laser-light Absorption and Back Reflection. *Physical Review Letters* 48 (15), 1018, 1982.

[64] Hoffman, J., Chrzanowska, J., Kucharski, S., Moscicki, T., Mihailescu, I. N., et al., The Effect of Laser Wavelength on the Ablation Rate of Carbon. *Applied Physics A* 117 (1), 395–400, 2014.

[65] Born, M. and Wolf, E., Principles of Optics: Electromagnetic Theory of Propagation, Interference and Diffraction of Light. New York: Cambridge University Press, 1999.

[66] Regenfuss, P., Streek, A., Hartwig, L., Klötzer, S., Brabant, T., et al., Principles of Laser Micro Sintering. *Rapid PrototypingJournal* 13 (4), 204–212, 2007.

[67] Chung Ng, C., Savalani, M. and Chung Man, H., Fabrication of Magnesium Using Selective Laser Melting Technique *RapidPrototyping Journal* 17 (6), 479–490, 2011.

[68] Sahasrabudhe, H. and Bandyopadhyay, A., AdditiveManufacturing of Reactive In Situ Zr Based Ultra-High Temperature Ceramic Composites. *JOM* 68 (3), 1–9, 2016.

[69] Ke, L., Zhu, H., Yin, J. and Wang, X., Effects of Peak Laser Poweron Laser Micro Sintering of Nickel Powder by Pulsed Nd: YAGLaser. *Rapid Prototyping Journal* 20 (4), 328–335, 2014.

[70] Agarwala, M., Bourell, D., Beaman, J., Marcus, H. and Barlow, J., Direct Selective Laser Sintering of Metals. *Rapid PrototypingJournal* 1 (1), 26–36, 1995.

[71] Perinchery, S., Smits, E., Sridhar, A., Albert, P., van den Brand, J., Mandamparambil, R. et al., Investigation of the Effects of LIFT Printing with a KrF-excimer Laser on Thermally Sensitive Electrically Conductive Adhesives. *Laser Physics* 24(6), 066101, 2014.

[72] Loy, J. and Tatham, P., Redesigning Production Systems In: *Handbook of Sustainability in Additive Manufacturing*, Springer, 2016.

[73] Mahamood, R. M. and Akinlabi, E. T., Laser Additive Manufacturing. *Advanced Manufacturing Techniques Using Laser Material Processing* 1, 2016.

[74] Lin, D., Nian, Q., Deng, B., Jin, S., Hu, Y., Wang, W. et al., Three-Dimensional Printing of Complex Structures: Man Made or toward Nature? *ACS Nano* 8(10), 9710–9715, 2014.

[75] Birtchnell, T. and Urry, J., *A New Industrial Future?: 3D Printing and the Reconfiguring of Production, Distribution, and Consumption*, Routledge, 2016.

[76] Chua, C. K. and Leong, K. F., *3D Printing and Additive Manufacturing: Principles and Applications*, Singapore: World Scientific, 2014.

[77] Chua, C., Chou, S. and Wong, T., A Study of the State-of-the-Art Rapid Prototyping Technologies. *The International Journal of Advanced Manufacturing Technology* 14(2), 146–152, 1998.

[78] Naing, M., Chua, C., Leong, K. and Wang, Y., Fabrication of Customised Scaffolds Using Computer-Aided Design and Rapid Prototyping Techniques. *Rapid Prototyping Journal* 11(4), 249–259, 2005.

[79] Thompson, S. M., Bian, L., Shamsaei, N. and Yadollahi, A., An Overview of Direct Laser Deposition for Additive Manufacturing; Part I: Transport Phenomena, Modeling and Diagnostics. *Addit. Manuf.* 8, 36–62, 2015.

[80] Dobbelstein, H., Gurevich, E. L., George, E. P., Ostendorf, A. and Laplanche, G., Laser Metal Deposition of Compositionally Graded TiZrNbTa Refractory High-Entropy Alloys Using Elemental Powder Blends. *Addit. Manuf.* 25, 252–262, 2018.

[81] Zhou, Y. H., Zhang, Z. H., Wang, Y. P., Liu, G., Zhou, S. Y., Li, Y. L., et al., Selective Laser Melting of Typical Metallic Materials: An Effective Process Prediction Model Developed by Energy Absorption and Consumption Analysis. *Addit Manuf.* 25, 204–217, 2018.

[82] Bourell, D. L., Perspectives on Additive Manufacturing. *Annual Review of Materials Research* (0), 2016.

[83] Wimpenny, D. I., Pandey, P. M. and Kumar, L. J., *[Advances in 3D Printing & Additive Manufacturing Technologies]*, Springer, 2016.

[84] Chua, C. K. and Yeong, W. Y., Bioprinting: Principles and Applications, *World Scientific*, 2014.

[85] Khoo, Z. X., Teoh, J. E. M., Liu, Y., Chua, C. K., Yang, S., An, J. et al., 3D Printing of Smart Materials: A Review on Recent Progresses in 4D Printing. *Virtual and Physical Prototyping* 10 (3), 103–122, 2015.

[86] Hull, C. W., Apparatus for Production of Three-dimensional Objects by Stereolithography, U.S. Patent No. 4,575,330, 1986.

[87] Lalevée, J., Blanchard, N., Tehfe, M. A., Peter, M., Morlet-Savary, F., et al., Efficient Dual Radical/cationic Photoinitiator under Visible Light: A New Concept. *Polymer Chemistry* 2 (9), 1986–1991, 2011.

[88] Decker, C., Kinetic Study of Light-induced Polymerization by Real-time UV and IR Spectroscopy. *Journal of Polymer SciencePart A: Polymer Chemistry* 30 (5), 913–928, 1992.

[89] Partanen, J., Solid State Lasers for Stereolithography. 7th AnnualSolid Freeform Fabrication Symposium Proceedings, 369–376,1996.

[90] Jacobs, F. P., Rapid Prototyping & Manufacturing: Fundamentals of Stereolithography. *Society of Manufacturing Engineers*, 1992.

[91] Lee, I. H. and Cho, D. W., Micro-stereolithography Photopolymer Solidification Patterns for Various Laser Beam Exposure Conditions. *The International Journal of Advanced Manufacturing Technology* 22 (5–6), 410–416, 2003.

[92] Stampfl, J., Baudis, S., Heller, C., Liska, R., Neumeister, A., et al., Photopolymers with Tunable Mechanical Properties Processed by Laser-based High-resolution Stereolithography. *Journal of Micromechanics and Microengineering* 18 (12), 125014, 2008.

[93] Zheng, X., Deotte, J., Alonso, M. P., Farquar, G. R., Weisgraber, T. H., et al., Design and Optimization of a Light-emitting Diode Projection Micro-stereolithography Three-dimensional Manufacturing system. *Review of Scientific Instruments* 83 (12), 125001, 2012

[94] Baldacchini, T., "Three-dimensional Microfabrication Using Two-Photon Polymerization: Fundamentals, Technology, and Applications," William Andrew, 2015

[95] Corbel, S., Dufaud, O. and Roques-Carmes, T., Materials for Stereolithography. *Stereolithography*, Springer US, 141–159, 2011

[96] Beaman, J. J. and Deckard, C. R., "Selective Laser Sintering with Assisted Powder Handling," U.S. Patent No. 4,938,816, Washington, DC: U.S, Patent and Trademark Office, 1990.

[97] Khaing, M. W., Fuh, J. Y. H. and Lu, L., Direct Metal Laser Sintering for Rapid Tooling: Processing and Characterisation of EOS Parts. *Journal of Materials Processing Technology* 113 (3), 269–272, 2001.

[98] Kruth, J. P., Wang, X., Laoui, T. and Froyen, L., Lasers and Materials in Selective Laser Sintering. *Assembly Automation* 23 (4), 357–371, 2003.

[99] Nelson, J. C., Selective Laser Sintering: A Definition of the Process and an Empirical Sintering Model. *UMI*, 1993.

[100] Williams, J. D. and Deckard, C. R., Advances in Modeling the Effects of Selected Parameters on the SLS Process. *Rapid Prototyping Journal* 4 (2), 90–100, 1998.

[101] Meiners, W., Wissenbach, K. and Gasser, A., Shaped Body Especially Prototype or Replacement Part Production. DE Patent 19, 1998.

[102] Kruth, J. P., Vandenbroucke, B., Vaerenbergh, V. J. and Mercelis, P., Benchmarking of Different SLS/SLM Processes as Rapid Manufacturing Techniques, 2005.

[103] Crafer, R. and Oakley, P. J., *Laser Processing in Manufacturing*, Springer Science & Business Media, 1992.

[104] Dashchenko, A. I., *Manufacturing Technologies for Machines of the Future: 21st Century Technologies*, Springer Science & Business Media, 2012.

[105] Ahn, D. G., Direct Metal Additive Manufacturing Processes and Their Sustainable Applications for Green Technology: A Review. *International Journal of Precision Engineering and ManufacturingGreen Technology* 3 (4), 381–395, 2016.

[106] Khademzadeh, S., Parvin, N. and Bariani, P. F., Production of NiTi Alloy by Direct Metal Deposition of Mechanically Alloyed Powder Mixtures. *International Journal of Precision Engineering and Manufacturing* 16 (11), 2333–2338, 2015.

[107] Seidel, C., Anstaett, C., Horn, M. and Binder, M., Status quo der metallischenn multimaterialverarbeitung mittels Laserstrahlschmelzen Fachtagung Werkstoffe Und Additive Fertigung Conf. (OGM), 2018.

[108] Chivel, Y., New Approach to Multi-Material Processing in Selective Laser Melting. *Phys. Procedia* 83, 891–898, 2016.

[109] Woidasky, J., Recyclingtechnik—fachbuch für lehre undpraxis. *Chemie Ing. Tech.* 89, 346, 2017.

[110] Ullrich, H., *Mechanische Verfahrenstechnik: Berechnung Und Projektierung*, Berlin: Springer, 1967.

[111] Li, L., Syed, W. U. H. and Pinkerton, A. J., Rapid Additive Manufacturing of Functionally Graded Structures Using Simultaneous Wire and Powder Laser Deposition. *Virtual Phys. Prototyp.* 1, 217–225, 2006.

[112] Hiller, J. D. and Lipson, H. STL 2.0: A Proposal for a Universal Multi-Material Additive Manufacturing File Format Proc. of Solid Freeform Fabrication Symp. (Austin, TX, USA), 266–278, 2009.

[113] Müller, A., Roslyakova, I., Sprenger, M., Git, P., Rettig, R., Markl, M., Körner, C. and Singer, R. F., MultOpt++: A Fast Regression-Based Model for the Development of Compositions With High Robustness Against Scatter of Element Concentrations. *Model. Simul. Mater. Sci. Eng.* 27, 024001, 2019.

[114] Gangwar, A. K. S., Rao, P. S. and Kumar, A., Bio-Mechanical Design and Analysis of Femur Bone. *Materials Today: Proceedings* 44 (Part 1), 2179–2187, 2021, ISSN 2214-7853 10.1016/j.matpr.2020.12.282.

[115] Gangwar, A. K. S., Rao, P. S., Kumar, A. and Patil, P. P., Design and Analysis of Femur Bone: BioMechanical Aspects. *Journal of Critical Reviews* 6 (4), 133–139, 2019, ISSN-2394-5125.

[116] Kumar, A., Mamgain, D. P., Jaiswal, H. and Patil, P., Modal Analysis of Hand Arm Vibration (Humerus Bone) for Biodynamic Response Using Varying Boundary Conditions Based on FEA. *Advances in Intelligent Systems and Computing* 308, 169–176, 2015. 10.1007/978-81-322-2012-1_18.

[117] Kumar, A., Behmad, S. I. and Patil, P., Vibration Characterization and Static Analysis of Cortical Bone Fracture Based on Finite Element Analysis. *Engineering and Automation Problems*, No 3-2014, 115–119, 2014. UDC- 621.

[118] Kumar, A., Jaiswal, H., Garg, T. and Patil, P., Free Vibration Modes Analysis of Femur Bone Fracture using varying boundary conditions based on FEA. *Procedia Materials Science* 6, 1593–1599, 2014. 10.1016/j.mspro.2014.07.142.

[119] Kumar, A., Gori, Y., Rana, S., Sharma, N. K. and Yadav, B., *FEA of Humerus Bone Fracture and Healing. Advanced Materials for Biomechanical Applications* (1st ed.) CRC Press, 2022, 10.1201/9781003286806.

[120] Kumar, A., Datta, A. and Kumar, A., Recent Advancements and Future Trends in Next-Generation Materials for Biomedical Applications. In *Advanced Materials for Biomedical Applications*; Kumar, A., Gori, Y., Kumar, A., Meena, C. S., Dutt, N., Eds.; Boca Raton, FL, USA: CRC Press, Chapter 1; 1–19, 2022.

[121] Prasad, A., Chakraborty, G. and Kumar, A. Bio-based Environmentally Benign Polymeric Resorbable Materials for Orthopedic Fixation Applications. In *Advanced Materials for Biomedical Applications*; Kumar, A., Gori, Y., Kumar, A., Meena, C. S., Dutt, N., Eds.; Boca Raton, FL, USA: CRC Press, Chapter 15; 251–266, 2022.

[122] Datta, A., Kumar, A., Kumar, A., Kumar, A. and Singh, V. P., Advanced Materials in Biological Implants and Surgical Tools. In *Advanced Materials for Biomedical Applications*; Kumar, A., Gori, Y., Kumar, A., Meena, C. S., Dutt, N., Eds.; Boca Raton, FL, USA: CRC Press, Chapter 2; 21–43, 2022.

[123] Kumar, A., Gangwar, A. K. S., Kumar, A., Meena, C. S., Singh, V. P., Dutt, N., Prasad, A. and Gori, Y., Biomedical Study of Femur Bone Fracture and healing. In *Advanced Materials for Biomedical Applications*; Kumar, A., Gori, Y., Kumar, A., Meena, C. S., Dutt, N., Eds.; Boca Raton, FL, USA: CRC Press, Chapter 14; 235–250, 2022.

[124] Patil, P. P., Gori, Y., Kumar, A. and Tyagi, M., Experimental Analysis of Tribological Properties of Polyisobutylene Thickened Oil in Lubricated Contacts. *Tribology International*. 159, 106983, 2021.

[125] Kumar, A., Rana, S., Gori, Y. and Sharma, N. K., Thermal Contact Conductance Prediction Using FEM Based Computational Techniques. In *Advanced Computational Methods in Mechanical and Materials Engineering*, Boca Raton, FL, USA: CRC Press, 183–220, 2021. ISBN 9781032052915.

[126] Kumar, A. and Patil, P. P., FEA Simulation and RSM based Parametric Optimization of Vibrating Transmission Gearbox Housing. *Perspect. Sci.* 8, 388–391, 2016.

[127] Patil, P. P., Sharma, S. C., Jaiswal, H. and Kumar, A., Modeling Influence of Tube Material on Vibration based EMMFS using ANFIS. *Procedia Mater. Sci.* 6, 1097–1103, 2014.

[128] Bhoi, S., Kumar, A., Prasad, A., Meena, C. S., Sarkar, R. B., Mahto, B. and Ghosh, A., Performance Evaluation of Different Coating Materials in Delamination for Micro-Milling Applications on High-Speed Steel Substrate. *Micromachines* 13, 1277, 2022. 10.3390/mi13081277

[129] Bhoi, S., Prasad, A., Kumar, A., Sarkar, R. B., Mahto, B., Meena, C. S. and Pandey, C., Experimental Study to Evaluate the Wear Performance of UHMWPE and XLPE Material for Orthopedics Application. *Bioengineering* 9, 676, 2022. 10.3390/bioengineering9110676

Chapter 2

Physics of Laser–Matter Interaction in Laser-Based Manufacturing

Abhishek Kumar, Aayush Pathak, Avinash Kumar, and Ashwani Kumar

CONTENTS

2.1 Terminology..45
2.2 Energy Levels and Transitions 46
    2.2.1 Electronic Levels..46
    2.2.2 Vibrational Levels 46
    2.2.3 Rotational Levels...48
2.3 Excitation..48
    2.3.1 Stimulated and Spontaneous Emission 48
2.4 Population Inversions..49
2.5 Amplification and Oscillation 50
References...53

2.1 TERMINOLOGY

The term "quantum physics" is often used as a formal name to designate laser physics. Its root dates to maser research in past where it was used frequently in academic research. The laser physics terms are mostly derived from spectroscopy and the underlying physics of atoms and molecules. It is important to mention a few important terms here [1–5]:

Energy level: It is defined as the quantum state of an atom or molecule.

Ground energy level: It is defined as the lowest possible energy level of an atom or molecule, and it is assumed to possess zero energy. The ground (energy) level is considered as the reference to measure the relative energy of any particular energy level.

Species: It is used to designate "atom or molecule" in case of studies related to the energy levels in the general term.

Transition: When a species undoes shift from one energy level to another, then this process is known as transition.

DOI: 10.1201/9781003402398-2

Excitation: When a species shifts from a lower energy level to a relatively higher energy level, then this phenomenon is known as excitation. The new state is called excited state, and the species require to absorb the energy equal to the difference in the energy between both the levels to attain it.

Deexcitation or decay: When a species shifts from a higher energy level to a lower energy level, then this process is known as deexcitation or decay. Decay generally occurs with the release of transition energy.

Photons: Photon is defined as a quanta of electromagnetic radiation. The energy interaction with the species takes place in the form of photons.

Light: With reference to the laser technology, the term "light" is used to represent electromagnetic radiation for visible light, infrared and ultraviolet parts of spectrum. Light has dual behavior, and it acts both as a particle (photon) as well as a wave. The nature of light can be classified based on three important parameters: (i) wavelength, (ii) frequency of oscillation of a wave, and (iii) energy of a photon.

2.2 ENERGY LEVELS AND TRANSITIONS

The laser physics generally deals with three types of energy levels.

2.2.1 Electronic Levels

It comes into the picture due to the electron configuration in a species. It can be understood using Bohr's model for hydrogen atom. In that case, a single electron gradually moves away from the nucleus when it gains energy, and eventually escapes the nucleus once it crosses the threshold energy barrier. The electronic system becomes complex when more atoms are added to the species which leads to various interactions generating multiple electronic energy levels [6–9].

Note:
- The electronic transitions in laser physics normally occur because of the electrons present in the outer shells. The energy associated with them is typically on the order of an electronvolt which corresponds to the visible light, near-infrared, and ultraviolet wavelengths.
- The inner-shell electronic transitions which happen closer to the nucleus involve relatively higher energies, which correspond to far-ultraviolet or X-ray photons, and hence are less favorable for laser action (Figure 2.1).

2.2.2 Vibrational Levels

It is generated by the vibration of the atoms causing harmonic motion in a molecule. Vibrations are quantized based on the possible vibration pattern.

LOWER LEVEL   **UPPER LEVEL**

NUCLEUS   ELECTRON

(a)

(b)

(c)

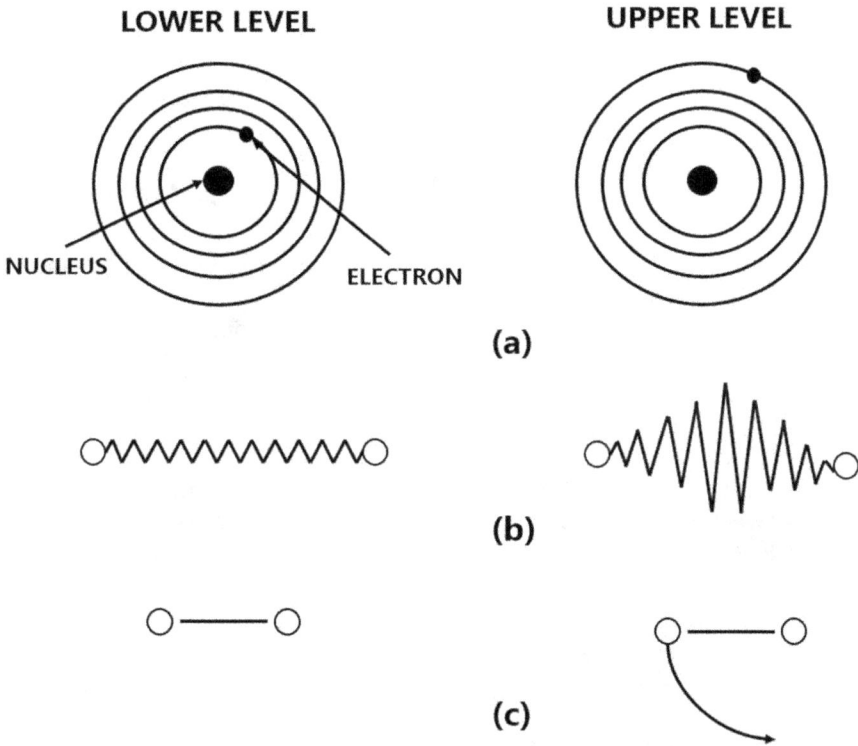

Figure 2.1 Important energy level transitions. (a) Electronic transitions (takes place in atoms and molecules); (b) vibrational; and (c) rotational transitions (takes place only in molecules) [1].

This creates multiple vibration levels corresponding to each vibration pattern (Figure 2.2). The vibration patterns increase with the number of atoms involved in a molecule. The vibrational transitions may be induced by shifting from one vibrational pattern to another, and usually correspond to infrared wavelengths (energies equivalent to the fraction of an electronvolt) [10–12].

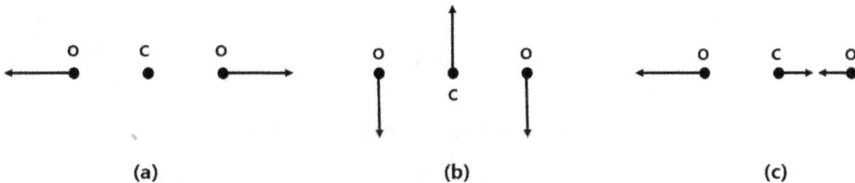

o   c   o   o   o   o   c   o

c

(a)   (b)   (c)

Figure 2.2 The vibrational modes in the case of CO_2 molecule. (a) Symmetric stretching (v_1 mode); (b) bending (v_2 mode); (c) asymmetric stretching (v_3 mode) [1].

2.2.3 Rotational Levels

It involves the quantized rotation of molecules and typically corresponds to far-infrared or sub millimeter wavelengths (energies are usually from 0.001 to 0.1 eV).

Note:
- Even though the various states of species such as electronic, vibrational, and rotational states are independent, still they may get altered simultaneously in a single transition.
- Atoms and ions don't have vibrational or rotational states, and hence their electronic transitions take place independently. Contrary to it, the changes in the vibrational state of molecules are generally complimented by the changes in the rotational states.
- Stark contrast is observed between the transitions involving the shifts in only one kind of energy level to that having simultaneous shifts between two or more types. The former is defined at a specific wavelength whereas the former transition is found to spread out over a range of wavelengths.

2.3 EXCITATION

It is one of the prerequisites for the laser action and is defined as the phenomenon of raising a species from a lower energy level to a higher one. Several different mechanisms can be used for excitation such as (a) photon absorption, (b) electron or ion collisions with species in the active medium, (c) atom and molecule collisions in the active medium, (d) recombination of free electrons with ionized atoms, (e) recombination of current carriers in a semiconductor, (f) chemical reactions producing excited species, and (g) acceleration of electrons [13–16].

Note:
- Absorption cross-section is often used to estimate the probability of a species to absorb energy for some of the above-mentioned mechanisms, and it depends on various factors such as the speed of the incident electrons, the wavelength of the illuminating light, and coincidences in the energy level structures of species.
- The laser excitation is a multi-step process.

2.3.1 Stimulated and Spontaneous Emission

Excited species can undergo excess energy release through nonradioactive processes such as emitting a photon or collisions with other atoms or molecules. The photon emission may be spontaneous or stimulated and occurs without outer intervention when a species experiences an energy drop to a lower level after a natural decay time (which is similar to the half-life of radioactive isotopes) (fraction of a second).

Note:

- Photon emission and transition to a lower energy state may also be stimulated by the presence of a photon having the same energy as the transition.
- It happens when there is a ready source of such photons and spontaneous emission by the same species is caused.
- The stimulated emission of photons is triggered by the first few spontaneously emitted photons, which leads to a cascade of stimulated emissions.
- The wavelength of stimulated emission is identical to that of the original photon. Also, the stimulated emission and the original light are in phase (coherent) with each other.

Spontaneous emission is responsible for the light from the celestial bodies such as the sun and stars as well as the normal objects like light bulbs, flames, etc. Stimulated emission generates laser light but sometimes its intensity would be too small to detect in practical life [17–23].

2.4 POPULATION INVERSIONS

It is scarce to observe stimulated emissions since the species would most likely plunge to the lowest available energy level. In case of the normal conditions (under thermodynamic equilibrium), the population of state is supposed to decrease with rise in energy (Figure 2.3). This implies the

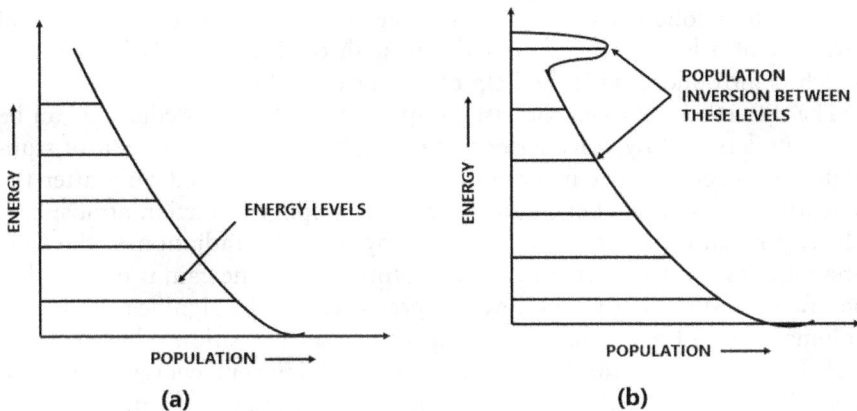

Figure 2.3 (a) Populations distribution among energy levels under equilibrium and (b) when population inversion takes place [1].

presence of larger population in the lower state of a transition when compared to that in the higher state, and hence a photon has more chances for getting absorbed by a lower-state species in place of stimulating emission from one in the higher state. This explains the reason for the dominance of spontaneous emission.

A population inversion is necessary to cause stimulated emission, i.e., the population of the upper level of a transition needs to be higher than that of the lower level. It will then lead a photon of the transition energy to stimulate emission from the excited state and reduces the possibility of its absorption by the lower state. It will then eventually result in laser gain or amplification, i.e., a rise in the number of photons with the transition energy which corresponds to the difference between the stimulated emission and absorption at that wavelength.

Note:
- The gain is typically measured as the percentage increase per pass or per centimeter of distance through the laser medium.
- The gain is sometimes also measured as the number of added photons generated for each centimeter of laser medium.
- The laser gain will be proportional to the difference in the chance of stimulated emission and the corresponding chance of absorption. It explains the importance of both the upper levels and lower levels of the laser transitions. Hence, the lower level must be vacated when the upper level is filled to maintain the laser action else the population inversion will end, thereby stopping the laser action.

2.5 AMPLIFICATION AND OSCILLATION

Laser light can be produced solely by the population inversion but it only results into a coherent light which is monochromatic in nature. The useful property of a laser is incorporated into it through the oscillation of light which is introduced with the help of a laser cavity.

The difference between the laser amplification and its oscillation can be better understood by considering the examples. The amplification of stimulated emission can occur naturally, though it was noticed only after the invention of the laser. Let us consider the example of Martian atmosphere [2]. A population inversion is produced by the solar radiation in the CO_2 present in the upper layers of the Mars atmosphere. The gain is observed to be low corresponding to the low gas pressure, but the significantly higher volume involved generates a 10.4-μm CO_2 laser transition which is quite high by the human standards. However, the inefficient energy extraction from those CO_2 molecules causes the laser emission intensity deterioration resulting into its dissipation in space as shown in Figure 2.4 which leads to a low intensity of those emissions reaching the earth and hence its existence

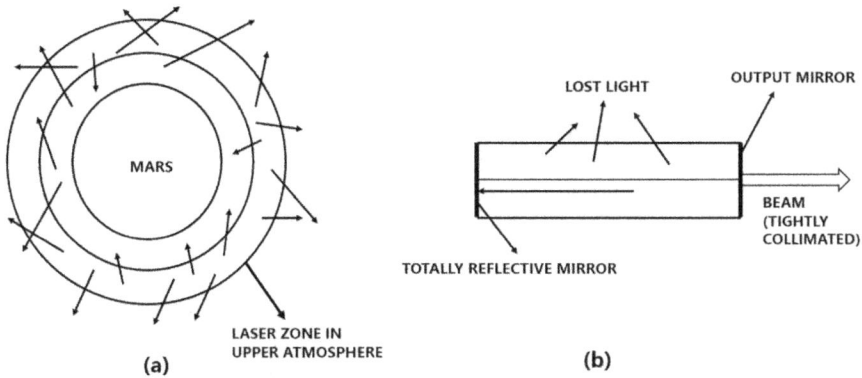

Figure 2.4 Directional dependence of laser beam. (a) The schematic shows the Martian atmosphere [2] and (b) beam generation using a resonant cavity [1].

was discovered only after the usage of sophisticated spectroscopic instruments. Contrary to the last example, an ordinary CO_2 laser works at relatively higher pressures but uses a comparatively much smaller volume of gas and still manages to extract energy more efficiently and successfully concentrates it into a narrow beam with the help of a pair of mirrors on either end of a cylindrical tube. If in case stimulated emission takes place on the axis between the two mirrors, then it is reflected back and forth through the tube which helps to stimulate emission continuously from CO_2 molecules. The stimulated emission is suppressed in other directions out of the laser medium that results in a concentrated beam of stimulated emission oscillating back and forth between the mirrors.

From the electronic perspective, the mirrors are used to obtain positive feedback. Theoretically, a pair of totally reflective mirrors can be used to amplify the laser beam to infinity, but the degree of amplification is limited in practice by the cavity losses. The cavity losses may occur either around the edge, through a partially transparent section, or through a hole in either of the mirror which lets a fraction of light escape from the laser cavity. This leaked light forms a laser beam, and it depends on the gain of the laser medium. In the case of an operating continuous-output laser, the total gain is the sum of the cavity losses and the fraction of energy allowed out of the cavity. Hence, the fraction of light transmitted out of the cavity is proportional to the gain. As an example, the gain in case of a helium–neon laser is low and hence the output mirrors allow only 1% of the incident light to escape as the laser beam. The higher gain means that the transmission will be higher. For example, in the case of excimer and nitrogen lasers, the gain is so high that pair of cavity mirrors are not required (it operates in super radiant mode).

Laser oscillation can also be understood as a threshold phenomenon that is caused when the laser medium exceeds the sum of cavity losses and output-mirror transmission. In that case, the gain is proportional to the

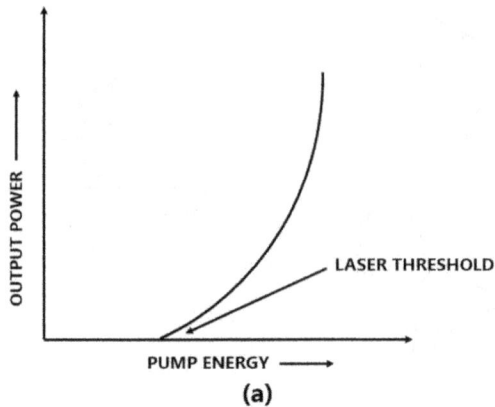

Figure 2.5 The phenomenon of Laser threshold [1].

operating conditions including the energy input to the laser medium. With the increase in input power, the laser output is zero till the threshold is passed and then it starts increasing steeply as shown in Figure 2.5. The threshold value is dependent on the laser-medium conditions and the cavity configurations. This technique is used to practically generate different power levels in lasers.

Note:
- The transparency of the output mirror dictates the power inside the laser cavity and that of the laser beam.
- When the transparency of the mirror is kept lower such as in the case of low gain media, the power levels inside the cavity is relatively much higher than that outside.
- In high-power applications, it is recommended to put optical accessories inside the laser cavity to take advantage of the higher powers available there. Hence, few lasers are designed with intracavity space to allow the placement of optics or samples inside the cavity.
- Most lasers are oscillators that include cavities defined by a pair of reflectors except the ones that have high gain and operate in super radiant mode.

Laser technologies providing methodologies for sustainable manufacturing and significant versatility in terms of materials to be converted into final product as well as the structures that can be tailored [17–23]. Along with more advanced applications from mechanical to electronic to biomedical industry require more complicated operations for precise production [24–26]. These developments have promoted laser-based technologies as they provide benefits in terms of technology, reliability, and the economy [27,28].

REFERENCES

[1] J. Hecht, *The Laser Guidebook, 2nd edition*. McGraw Hill, 1999, ISBN-10: 0071359672.

[2] M. Mumma, et al., "Discovery of natural gain amplification in the 10-micrometer carbon-dioxide laser bands on Mars: A natural laser," *Science*, vol. 212, pp. 45–50, 1981.

[3] J. Peal, I. Matthews, C. Runnett, H. Thomas, D. Ripley, "An update on cardiac implantable electronic devices for the general physician," *Journal of the Royal College of Physicians of Edinburgh*, vol. 48, no. 2, pp. 141–147, 2018.

[4] L. Barbosa, R. Jordan, J. Cordioli, "Structural dynamic analysis of a BTE hearing aid case," in *7th National Congress of Mechanic Engineering—CONEM*, 2012.

[5] S. Nisar, M. Sheikh, L. Li, S. Safdar, "Effect of thermal stresses on chipfree diode laser cutting of glass," *Optics & Laser Technology*, vol. 41, no. 3, pp. 318–327, 2009.

[6] J. R. Lawrence, *Advances in Laser Materials Processing: Technology, Research and Applications*. Woodhead Publishing, 2017.

[7] L. Li, C. Diver, J. Atkinson, R. Giedl-Wagner, H. Helml, "Sequential laser and EDM micro-drilling for next generation fuel injection nozzle manufacture," *CIRP Annals*, vol. 55, no. 1, pp. 179–182, 2006.

[8] C. Tan, F. Weng, S. Sui, Y. Chew, G. Bi, "Progress and perspectives in laser additive manufacturing of key aeroengine materials," *International Journal of Machine Tools and Manufacture*, vol. 170, p. 103804, 2021, ISSN: 0890-6955. https://doi.org/10.1016/j.ijmachtools.2021.103804 [Online]. Available: https://www.sciencedirect.com/science/article/pii/S0890695521001139.

[9] T. Schopphoven, A. Gasser, K. Wissenbach, R. Poprawe, "Investigations on ultra-high-speed laser material deposition as alternative for hard chrome plating and thermal spraying," *Journal of Laser Applications*, vol. 28, no. 2, p. 022501, 2016.

[10] I. Mingareev, M. Richardson, "Laser additive manufacturing: Going mainstream," *Optics and Photonics News*, vol. 28, no. 2, pp. 24–31, 2017.

[11] A. R. Choudhury, T. Ezz, L. Li, "Synthesis of hard nano-structured metal matrix composite boride coatings using combined laser and sol–gel technology," *Materials Science and Engineering: A*, vol. 445, pp. 193–202, 2007.

[12] B. Vayre, F. Vignat, F. Villeneuve, "Designing for additive manufacturing," *Procedia CIRP*, vol. 3, pp. 632–637, 2012, 45th CIRP Conference on Manufacturing Systems 2012, ISSN: 2212-8271. https://doi.org/10.1016/j.procir.2012.07.108 [Online]. Available: https://www.sciencedirect.com/science/article/pii/S2212827112002806.

[13] G. Zhu, Z. Xu, Y. Jin, et al., "Mechanism and application of laser cleaning: A review," *Optics and Lasers in Engineering*, vol. 157, p. 107130, 2022, ISSN: 0143-8166. https://doi.org/10.1016/j.optlaseng.2022.107130 [Online]. Available: https://www.sciencedirect.com/science/article/pii/S0143816622001828.

[14] S. Bhoi, A. Prasad, A. Kumar, R. B. Sarkar, B. Mahto, C. S. Meena, "Experimental study to evaluate the wear performance of UHMWPE and XLPE material for orthopedics application," *Bioengineering*, vol. 9, p. 676, 2022. 10.3390/bioengineering9110676

[15] V. P. Singh, S. Jain, A. Karn, A. Kumar, G. Dwivedi, C. S. Meena, R. Cozzolino, "Mathematical modeling of efficiency evaluation of double-pass parallel flow solar air heater," *Sustainability*, vol. 14, p. 10535, 2022. 10.33 90/su141710535

[16] S. Bhoi, A. Kumar, A. Prasad, C. S. Meena, R. B. Sarkar, B. Mahto, A. Ghosh, "Performance evaluation of different coating materials in delamination for micro- milling applications on high-speed steel substrate," *Micromachines*, vol. 13, p. 1277, 2022. 10.3390/mi13081277

[17] A. K. S. Gangwar, P. S. Rao, A. Kumar, "Bio-Mechanical Design and Analysis of Femur Bone," *Materials Today: Proceedings*, vol. 44, no. Part 1, pp. 2179–2187, 2021, ISSN 2214-7853. 10.1016/j.matpr.2020.12.282

[18] A. Kumar, S. Gautam, Y. Gori, P. Patil, "Dissimilar materials welded specimen analysis using FEA," *Journal of Critical Reviews*, vol. 6, no. 5, pp. 356–362, 2019, ISSN-2394-5125.

[19] A. Kumar, P. P. Patil, "Modal analysis of heavy vehicle truck transmission gearbox housing made from different materials," *Journal of Engineering Science and Technology*, vol. 11, no. 2, pp. 252–266, 2016.

[20] A. Kumar, H. Jaiswal, R. Jain, P. P. Patil, "Free vibration and material mechanical properties influence based frequency and mode shape analysis of transmission gearbox casing," *Procedia Engineering*, vol. 97, pp. 1097–1106, 2014. 10.1016/j.proeng.2014.12.388

[21] A. Kumar, S. I. Behmad, P. P. Patil, "Vibration characterization and static analysis of cortical bone fracture based on finite element analysis," *Engineering and Automation Problems*, no. 3, pp. 115–119, 2014. UDC- 621.

[22] P. P. Patil, S. C. Sharma, H. Jaiswal, A. Kumar, "Modeling influence of tube material on vibration based EMMFS using ANFIS," *Procedia Materials Science*, vol. 6, pp. 1097–1103, 2014. 10.1016/j.mspro.2014.07.181

[23] P. Patil, S. Sharma, A. Saini, A. Kumar, "ANN modelling of Cu type Omega vibration based mass flow sensor," *Procedia Technology*, vol. 14, pp. 260–265, 2014. 10.1016/j.protcy.2014.08.034

[24] A. Prasad, G. Chakraborty, A. Kumar, "Bio-based environmentally benign polymeric resorbable materials for orthopedic fixation applications," *Advanced Materials for Biomedical Applications*. CRC Press, Taylor & Francis, 2022. ISBN: 9781003344810, 10.1201/9781003344810-15

[25] S. Singh, A. Kumar, S. K. Behura, K. Verma, "Challenges and opportunities in nanomanufacturing," *Nanomanufacturing and Nanomaterials Design: Principles and Applications*. CRC Press, Taylor & Francis, 2022. ISBN: 9781003220602. 10.1201/9781003220602-2

[26] S. Srivastava, D. Verma, S. Thusoo, A. Kumar, V. P. Singh, R. Kumar, "Nanomanufacturing for energy conversion and storage devices," *Nanomanufacturing and Nanomaterials Design: Principles and Applications*. CRC Press, Taylor & Francis, 2022, ISBN: 9781003220602. 10.1201/9781003220602-10

[27] G. Chakraborty, A. Prasad, A. Kumar, "Processing of biodegradable composites," *Biodegradable Composites for Packaging Applications*. CRC Press, Taylor & Francis, 2022, ISBN: 978103227908. 10.1201/9781003227908-3

[28] A. Kumar, Y. Gori, S. Rana, N. K. Sharma, B. Yadav, "FEA of humerus bone fracture and healing," *Advanced Materials for Biomechanical Applications*. CRC Press (Taylor and Francis), 2022, ISBN: 9781032054490. 10.1201/ 9781003286806-14

Chapter 3

Current Issues, Developments, and Constraints in Additive Manufacturing

Sushma Katti, V. S. Patil, G. H. Pujar, H.J. Amith Yadav, and M. N. Kalasad

CONTENTS

3.1   Introduction 55
3.2   Fundamentals of AM 56
    3.2.1   Process.. 56
    3.2.2   Technique 57
    3.2.3   Materials 57
    3.2.4   Understanding the Requirements of Standards................... 57
3.3   Recent Trends 58
    3.3.1   Hybrid Process 58
    3.3.2   Micro Manufacturing... 58
    3.3.3   Process Optimization... 58
    3.3.4   4D Printing.. 58
    3.3.5   Control and Monitoring... 59
    3.3.6   Data Acquisition 59
    3.3.7   Sustainability 59
3.4   Applications 59
    3.4.1   Biomaterials... 59
    3.4.2   Aerospace 60
    3.4.3   Buildings... 60
    3.4.4   Protective Structures.. 60
    3.4.5   Orthopedics... 61
    3.4.6   Repair and Remanufacturing of Damaged Components 61
    3.4.7   AM for Space Resources 61
3.5   Limitations of AM 62
3.6   Conclusion 62
References... 63

3.1 INTRODUCTION

According to the definition of the American Society for Testing and Materials (ASTM), additive manufacturing (AM) is the process of joining

DOI: 10.1201/9781003402398-3

the material layer by layer using three-dimensional Computer-Aided Design (3D CAD). Basically, AM is a computer-controlled process. Different terminologies like 3D printing (3DP), solid freeform fabrication (SFF), rapid prototyping (RP), rapid manufacturing (RM), and direct digital manufacturing (DDM) are generally used to define AM processes [1]. The AM is also known as 3D printing, which plays a vital role in the advancement of technology. The first form of 3D printing was invented by Charles Chuck Hull in 1980 and it was called stereolithography (SLA). In this technique, photocurable polymer will be partially cured by focusing UV light on the polymer. Uncured polymer left in the bath and offers support to the part being constructed. The printed first layer moves down in the polymer solution and subsequent polymer layer will be available above the following layer. This procedure will be continued until the part is finished based on the CAD design. With the growing demands, interests, and popularity, AM is continuously customized, reimagined, and redefined by researchers in many fields [2]. AM promotes the production of high-quality designs of complex geometries and low production volumes, which are difficult to fabricate using conventional systems [3] because of its easy usage of AM has widespread industrial applications involving aerospace, medical, aviation, and automation sectors. So, it is considered to be a high-potential manufacturing method [4,5]. AM is very helpful for the manufacturing required frequent design changes. In this way, AM offers facility to produce complex parts by conquering the design constraints of conventional manufacturing processes. Therefore, this technique is an important resource tool that offers to create custom or sophisticated models in one step by overcoming the traditional manufacturing limitations such as the need for peculiar tooling, difficulty in manufacturing complex shapes, and high material waste [6].

Although AM has many benefits, most industrial organizations still do not consider AM as a feasible alternative to traditional manufacturing processes. Because of low precision and accuracy, poor productivity, and long build times. But, however, due to the rapid growth and enormous development of technology, it is likely considered to address the main barriers for the adoption of AM [7].

The present review covers the primary information like processes, materials, current issues, developments, and constraints of AM.

3.2 FUNDAMENTALS OF AM

3.2.1 Process

Using computer-aided design (CAD), AM takes into consideration the formation of items with exact mathematical shapes. These are fabricated layer by layer, likewise with a 3D printing process, which is as opposed to conventional assembling that frequently requires machining or different strategies to eliminate surplus material.

Classification of AM processes depends on various boundaries, for example, the kind of material utilized, hardening cycle of the material, and affidavit ways [8].

These cycles were recently ordered by various specialists [9–11], and have now been standardized by the ASTM Worldwide Board of trustees F42 on AM Technologies into the seven classes.

3.2.2 Technique

The part is manufactured by designing using CAD software and feeding the converted form into the printer's comprehensive format emphasizing the triangular facets of the model. Thereafter with the requisite conditions set up of printer, the part is built layer-by-layer deposition of material. Post-processing of part is exposed to application and innovation utilized [12].

3.2.3 Materials

The diverse materials used for AM. The kind of substance used prominently affects factors like shape, cost, and durability of a printed part, thus by limiting potential applications. In general, three main categories of materials, i.e., liquids, solids, and powder are used. Each of these three categories includes a variety of materials like ceramics, composites, metals, and polymers [13]. Depending on the type of process, appropriate materials should be chosen for the process [14].

3.2.4 Understanding the Requirements of Standards

The traditional principles utilized in the production of materials or innovation are not pertinent in AM [15]. Over years now, the requirement for new guidelines has been broadly perceived as a need in scholarly communities, ventures, and exploration associations. Generally believe that, by critically controlling the quality of manufactured parts AM can be applied to a larger extent in sectors of aerospace, medicine, etc. [16]. Many administering pathways made by analysts in the USA and EU, and reports distributed by the science and innovation strategy establishment likewise focused on the way that the absence of principles is hampering the creation in AM [17,18]. Considering the needs of the customer, recognizing the importance of standards on a national and international level upgrading the AM standards is essential.

Hence, Standardization, Innovation, and Research (STAIR) along with Support Action for Standardization in Additive Manufacturing (SASAM) organizations have taken initiative and are working continuously in developing standards of AM across the international community.

3.3 RECENT TRENDS

In this segment, probably the most recent patterns in AM production are momentarily talked about. Some significant exploration subjects are introduced to give a layout of the present status of the fundamental examination and improvements in AM.

3.3.1 Hybrid Process

Hybrid machines that have abilities of joining AM with regular advancements are being laid out somewhat recently [19]. This strategy utilizes an added substance cycle to fabricate a close net shape which will additionally handle its last shape subsequently by getting the ideal exactness [20]. Thus, the issue of AM on surface completion and accuracy is settled and, the capability of subtractive assembling is extended to parts of more complex calculations [21].

3.3.2 Micro Manufacturing

Since most of the features must be greater than 0.5 mm, AM takes place at macroscale levels. Selection of several methods appropriate for micro-AM process was done by Bhushan and Caspers [22], amongst them, inkjet and SLA processes have given the results with the smaller size. The obtained samples' thickness was 200 µm (SLA) and 340 µm (inkjet). The results obtained by performing micro extrusion in AM with high-aspect-ratio (HAR) nozzles showed that various extrusion parameters affect the quality of the deposited plane. These parameters include piston speed, the distance between the nozzle and the substrate, nozzle moving speed, extrusion rate, the expulsion postpone in light of the difference in the slam speed and air pockets caught inside the material repository [23].

3.3.3 Process Optimization

Various interaction boundaries impact the nature of the pieces acquired by AM. To get the fundamental surfaces, the cycle boundaries must be enhanced by making various blends in the underlying stage in the standard triangle language (STL) document, alluding to the direction of the piece, development direction and layer thickness, to acquire proficient result necessities [24].

3.3.4 4D Printing

The idea of 4D printing arose in 2013. 4D printing is the interaction through which a 3D-printed object changes itself into one more design over the impact of outside energy input such as temperature, light or other ecological boosts. It has properties of multifunctionality, self-gathering, and self-fixes. For designs to advance over the long run, they require extra upgrades and astute materials delicate to feeling [25].

3.3.5 Control and Monitoring

Quality assurance and control are the biggest challenges executed by AM. To overcome this, monitoring systems to improve the quality of the parts and inspection during AM is essential. According to Chua et al. [26] the size and temperature profile of the combination get together are basic issues in the observing and examination processes. Moreover, they likewise propose a review strategy and closed circuit monitoring system to meet quality control of AM for metals.

3.3.6 Data Acquisition

The required data is obtained by scanning through Computed Tomography (CT) and Magnetic Resonance Imaging (MRI) [27] and transferring data into CAD/STL files which are further printed. A study report presented by Manmadhachary [27] on AM of an adult human dry mandible says that it is feasible to get the scaled STL model to print the dry mandible utilizing FDM at last.

3.3.7 Sustainability

Currently, sustainability is one of the main concepts in manufacturing. Administrations, researchers, and companies are concentrating on sustainable production processes, which are environment-friendly. The adoption of AM technology is economical, but the influence of other factors such as social and sustainability are minimum [28]. One of the research findings suggests that the reuse of non-used plastics in production of powders for selective laser sintering to prepare feedstock for inexpensive high-value fused deposition modeling products, resulting in important energy savings and thus, reducing the environmental impact [29].

3.4 APPLICATIONS

3.4.1 Biomaterials

The biomedical market is one of the drivers for AM evolution and growth, representing 11% of the total AM market share today. Biomedical applications have interesting prerequisites like high intricacy, customization and patient-explicit necessities, little creation amounts, and simple free [30]. Novel AM drug conveyance frameworks are being created: solid dosage forms (the most read up for its simple commercialization) [31]; implantable drug delivery vehicles [32]; and topical drug delivery systems [33]. AM had some control over the delivery profile of medications by changing the 3D shape [31], the micro-architecture of drug delivery systems, and as well as the place of the dynamic specialists [34]. AM is additionally changing the embed business, i.e., it is presently conceivable to develop patient-explicit inserts [35].

3.4.2 Aerospace

Aviation application possesses 18.2% of the absolute AM market today and is viewed as one of the most encouraging fields from here on out. The unconventional attributes like complex calculation, challenging to-machine materials, redid creation, on-request manufacturing, high-execution to weight proportion, and high purchase-to-fly proportion makes AM procedures ideal for aviation parts. The assembling or fix of both metallic and non-metallic parts like air motor parts, turbine edges, and intensity exchangers should be possible by AM [36,37]. Non-metal AM techniques, for example, stereolithography, multi-fly demonstrating [38] and fused deposition modeling (FDM) are utilized for the quick prototyping of parts and for assembling apparatuses and interiors made of plastics, ceramics, and composite materials. Advancement of part plan and combination with different capabilities are finished by the high accuracy of powder bed fusion (PBF) This procedure is utilized primarily for more modest parts with higher intricacy. AM will likewise change the extra parts supply chains [39].

3.4.3 Buildings

Automated building construction with 3D printing technology has gained increasing attention in recent years. It has revolutionized the construction industry by offering astronauts easy construction on the moon [40]. It offers a critical decrease in development time and labor [41]. Many researchers described and summarized promising techniques like large-scale 3D desktop 3D printing techniques, etc., which can change the development business later on [42–44]. The tensile strength of the concrete can be increased by replacing steel reinforcement [45].

3.4.4 Protective Structures

Due to the great advances in the development of novel protective systems and rapid evolution of protective structures, many lattice structures and innovative solutions showing excellent results are developed, but some improvements are necessary.

Advances in AM and material creation advances like brilliant and lightweight designs with high stiffness-to-weight and high solidarity-to-weight proportions have augmented economic efficiency and efficiency of the protective system by reducing weight. These technologies are also used for manufacturing lattice structures [46].

3.4.5 Orthopedics

AM producing gives colossal advancement in producing field and presently investigate its applications in different medical fields. The 3D-printed model's first objective is to look at the cases in the facility and give a definite outline to specialists. The head utilization of this model is for the testing strategy in advance as it gives a sensation of the mechanical reaction of genuine issue that remains to be worked out for specialists. The activity can likewise be performed on the 3D model to look at and picture prior to performing a real medical procedure.

The uses of AM in orthopedic medical procedures give reproducible, protected, and solid models that work on quiet results and decreased working time as contrasted with traditional surgical strategies. This innovation can turn out effectively for preoperative preparation, schooling, and custom assembling. For custom assembling applications, it is utilized for prosthetics, careful aides, and inserts. 3D-printed models of life systems have aided the training of understudies, learners, patients, and specialists. Specialists can utilize the 3D-printed bone models for rehearsing a medical procedure and make sense of the patient or understudies about medical procedures. It additionally can possibly increment the quantity of apparatuses for the specialist and effectively embrace the upgrading. It turns into an important device that impacts each area of medication. Specialists can now foster new apparatuses and strategies for surgeries [47].

3.4.6 Repair and Remanufacturing of Damaged Components

In biomedical, AM is promising in part fix and remanufacturing, in which the harmed district of a part is topped off/fixed with filler materials to recuperate the missing geometry. The basic surfaces of the fixed parts can go through the last machining to get a surface completion. Fixing existing harmed parts is basic to boost their administration life and lessen the expenses of substitutions [48–56].

3.4.7 AM for Space Resources

AM and its related strategies could be consolidated along with the in situ asset usage (ISRU) idea, for building a scope of actual resources off-world. This could occur by successfully utilizing the bountiful and promptly accessible regular assets on location. Moreover, AM strategies enjoy the unmistakable benefit of being capable to work in an independent manner, without the requirement for ceaseless oversight or direct contribution from machine administrators. This capacity to work independently is one of the primary justifications for why AM is considered as being ideal for remote and brutal conditions, for example, the outer layer of planets like Mars or Moon. The materials for supporting such an office could be either privately exhumed regoliths straight forwardly, its beneficiated items like extricated metals, or results of the beneficiation processes that would regularly be futile for some other application [57,58].

3.5 LIMITATIONS OF AM

Taking in thought all introduced at this point it seems as though AM is nearly "all-powerful" innovation with basically no restrictions. This is a long way from the truth. This innovation request changes in the approach of part fabricating. In AM, the material isn't being taken off, however, added where is required. Creation in layers gives an open door to take off controls with respect to shape intricacy in huge percent; however, complete shortfall of machining process is problematic.

Some of AM's innovation constraints are as follows:

- Insignificant wall thickness
- Building volume is restricted agreeing machine
- Restricted measure of material powder types is accessible
- The cost of material is high because of the powder fabricating process cost
- Upholds are required in the assembling process
- Surface quality
- Leftover burdens

Notwithstanding the abovementioned, there are more constraints, and they can all cause troubles during the creation cycle or in double-dealing [59].

3.6 CONCLUSION

This chapter presents of latest trends and applications in AM. Initially, fundamentals covering process, technique, materials, and standards are introduced. Some of the recent trends and applications are discussed. Amongst them, it is possible to highlight the (i) research and development of Hybrid machines that join AM with traditional assembling processes are being finished. (ii) 4D printing which offers multifunctionality has the property of self-assembly and self-repair. Studies on micro-AM indicate that AM takes place at the macroscale level. (iii) The biomedical market is one of the drivers for AM evolution and growth. (iv) Aerospace application is viewed as quite possibly of the most encouraging field from here on out.

Promising techniques like large-scale 3D desktop 3D printing techniques, etc., can transform the construction industry in the future. (v) Advances in AM and material production technologies have augmented the economic efficiency and efficiency of the protective system.

REFERENCES

[1] Abdulhameed O., Al-Ahmari A., Ameen W., Mian S. H. Additive manufacturing: Challenges, trends, and applications. *Advances in Mechanical Engineering*. 2019, 11(2). 10.1177/1687814018822880

[2] Gao W., Zhang Y., Ramanujan D., et al. The status challenges and future of additive manufacturing in engineering. *Computer-Aided Design*. 2015, 69, 65–89.

[3] Vafadar A., Guzzomi F., Rassau A., Hayward K. Advances in metal additive manufacturing: A review of common processes, industrial applications, and current challenges. *Applied Science* 2021, 11, 1213. 10.3390/app11031213

[4] Singh S., Prakash C., Ramakrishna S. 3D printing of polyether-ether-ketone for biomedical applications, *European Polymer Journal*. 2019, 114(February), 234–248. 10.1016/j.eurpolymj.2019.02.035

[5] Babbar A., Jain V., Gupta D., Prakash C., Singh S., Sharma A. 3D bioprinting in pharmaceuticals, medicine, and tissue engineering applications. *Advanced Manufacturing and Processing Technology*. 2020, 147–161, 10.1201/9780429298042-7

[6] ISO; ASTM. Additive manufacturing—design—requirements, guidelines and recommendations. In *ISO/ASTM 52910:2018(E)*; Geneva, Switzerland: ISO; ASTM: West Conshohocken, PA, USA, 2018.

[7] Tofail S. A. M., Koumoulos E. P., Bandyopadhyay A., Bose S., O'Donoghue L., Charitidis C. Additive manufacturing: Scientific and technological challenges, market uptake and opportunities. *Materials Today*. 2018, 21-1, 22–37.

[8] Wong K. V., Hernandez A. A review of additive manufacturing. *ISRN Mechanical Engineering*. 2012, 1–10, 10.5402/2012/208760

[9] Hopkinson M. N., Gee A. D., Gouverneur V. Gold catalysis and fluorine. *Israel Journal of Chemistry*. 2010, 50, 675–690. 10.1002/ijch.201000078

[10] Gibson I., Rosen D. W., Stucker B. *Additive Manufacturing Technologies*. Springer. 2010. 10.1007/978-1-4419-1120-9.

[11] Hartke B. Non-deterministic global structure optimization: An introductory tutorial. *Reviews in Computational Chemistry*. 2022, 32, 1–43.

[12] Gibson I., Rosen D., Stucker B. (Eds.) *Additive Manufacturing Technologies*. New York, NY: Springer New York, 2015.

[13] Noorani, R. (Eds.) (2018). 3D Printing Technology, Applications, and Selection, ISBN 9780367781965, CRC Press: Boca Raton, FL, USA.

[14] Chua C. K., Wong C. H., Yeong W. Y. Chapter 1: Introduction to 3D printing or additive manufacturing. *Standards, Quality Control, and Measurement Sciences in 3D Printing and Additive Manufacturing*. 2017, 1–29.

[15] Kawalkar R., Kumar D. H., Lokhande S. P. A review for advancements in standardization for additive manufacturing. *Materials Today: Proceedings*. 2022, 50(Part 5), 1983–1990, ISSN 2214-7853, 10.1016/j.matpr.2021.09.333

[16] Kang H. S., Lee J. Y., Choi S., Kim H., Park J. H., Son J. Y., Kim B. H., Noh S. D. Smart manufacturing: Past research, present findings, and future directions. *International Journal of Precision Engineering Manufacturing – Green Technology*. 2016, 3(1), 111–128, 10.1007/s40684-016-0015-5

[17] Bourell D. L., Leu M. C., Rosen D. W. Roadmap for Additive *Manufacturing: Identifying the Future of Freeform Processing*. University of Texas at Austin Laboratory for Freeform Fabrication Advanced Manufacturing Center. 2009.

[18] Scott J., Gupta N., Weber C., Newsome S., Wohlers T., Caffrey T. *Additive manufacturing: Status and opportunities*. Science and Technology Policy Institute. 2012.

[19] Yamazaki T., *Development of A Hybrid Multi-tasking Machine Tool: Integration of Additive Manufacturing Technology with CNC Machining Procedia CIRP*. 2016, 42, 81–86

[20] Zhu Z., Dhokia V. G., Nassehi A., Newman S. T. A review of hybrid manufacturing processes – state of the art and future perspectives. *International Journal of Computer Integrated Manufacturing*. 2013, 26, 596–615.

[21] Sun Y.-J., Yan C., Wu S.-W., Gong H., Lee C.-H. Geometric simulation of 5-axis hybrid additive-subtractive manufacturing based on Tri-dexelmodel. *The International Journal of Advanced Manufacturing Technology*. 2018, 99, 2597–2610.

[22] Bhushan B., Caspers M. An overview of additive manufacturing (3D printing) for microfabrication. *Microsystem Technology*. 2017, 23, 1117–1124.

[23] Shaw L., Islam M., Li J., Li L., Ayub S. M. I. High-speed additive manufacturing through high-aspect-ratio nozzles. *JOM*. 2018, 70, 284–291.

[24] Kumbhar N. N., Mulay V. Post processing methods used to improve surface finish of products which are manufactured by additive manufacturing technologies: A review. *Journal of the Institution of Engineers (India): Series C*. 2016, 10.1007/s40032-016-0340-z.

[25] Momeni F., Hassani S. M. M., Liu X., Ni J. A review of 4D printing. *Materials and Design*. 2017, 122, 42–79.

[26] Chua Z. Y., Ahn I. H., Moon S. K. Process monitoring and inspection systems in metal additive manufacturing: Status and applications. *International Journal of Precision Engineering and Manufacturing-Green Technology*. 2017, 4, 235–245.

[27] Manmadhachary A. CT imaging parameters for precision models using additive manufacturing. Multiscale and multidisciplinary modeling. *Experiments and Design*. 2019, 2, 209–220.

[28] Niaki M. K., Torabi S. A., Nonino F. Why manufacturers adopt additive manufacturing technologies: The role of sustainability. *Journal of Cleaner Production*. 2019, 222, 381–392.

[29] Kumar S., Czekanski A. Roadmap to sustainable plastic additive manufacturing materials today. *Communications*. 2018, 15, 109–113.

[30] Ngo T. D., Kashani A., Imbalzano G., Nguyen K. T. Q., Hui D. Additive manufacturing (3D printing): A review of materials, methods, applications and challenges. *Composites Part B: Engineering*. 2018, 143, 172–196, ISSN 1359-8368, 10.1016/j.compositesb.2018.02.012.

[31] Goyanes A., Martinez P. R., Buanz A., Basit A. W., Gaisford S. Effect of geometry on drug release from 3D printed tablets. *International Journal of Pharmaceutics*. 2015, 494(2), 657–663.

[32] Huang W., Zheng Q., Sun W., Xu H., Yang X. Levofloxacin implants with predefined microstructure fabricated by three-dimensional printing technique. *International Journal of Pharmaceutics*. 2007, 339(1), 33–38.

[33] Goyanes A., Det-Amornrat U., Wang J., Basit A. W., Gaisford S. 3D scanning and 3D printing as innovative technologies for fabricating personalized topical drug delivery systems. *Journal of Controlled Release*. 2016, 234, 41–48.

[34] Goyanes A., Wang J., Buanz A., Martínez-Pacheco R. N., Telford R., Gaisford S., Basit A. W. 3D printing of medicines: Engineering novel oral devices with unique design and drug release characteristics. *Molecular Pharmaceutics*. 2015, 12(11), 4077–4084.

[35] Wang X., Xu S., Zhou S., Xu W., Leary M., Choong P., Qian M., Brandt M., Xie Y. M. Topological design and additive manufacturing of porous metals for bone scaffolds and orthopaedic implants: A review. *Biomaterials*. 2016, 83, 127–141.

[36] Yin L., Doyhamboure–Fouquet J., Tian X., Li D. Design and characterization of radar absorbing structure based on gradient-refractive-index metamaterials. *Composite B: Engineering*. 2018, 132, 178–187.

[37] Turco E., Golaszewski M., Giorgio I., D'Annibale F. Pantographic lattices with nonorthogonal fibres: Experiments and their numerical simulations. *Composite B: Engineering*. 2017, 118, 1–14.

[38] Chua C. K., Leong K. F., Lim C. S. *Rapid Prototyping: Principles and Applications*. World Scientific. 2010

[39] Liu P., Huang S. H., Mokasdar A., Zhou H., Hou L. The impact of additive manufacturing in the aircraft spare parts supply chain: Supply chain operation reference (SCOR) model based analysis. *Production Planning Control*. 2014, 25(13–14), 1169–1181.

[40] Labeaga-Martínez N., Sanjurjo-Rivo M., Díaz-Álvarez J., Martínez-Frías J. Additive manufacturing for a Moon village. *Procedia Manufacturing*. 2017, 13, 794–801.

[41] Wu P., Wang J., Wang X. A critical review of the use of 3-D printing in the construction industry. *Automation in Construction*. 2016, 68, 21–31.

[42] Lim S., Buswell R. A., Le T. T., Austin S. A., Gibb A. G. F., Thorpe T. Developments in construction-scale additive manufacturing processes. *Automation in Construction*. 2012, 21, 262–268.

[43] Nadal A. 3D printing for construction: A procedural and material-based approach. *Informes De La Construccion*. 2017, 69(546), e193.

[44] Hager I., Golonka A., Putanowicz R. 3D printing of buildings and building components as the future of sustainable construction? International conference on ecology and new building materials and products, ICEBMP 2016, May 31, 2016-June 2, 2016. Cerna Hora, Czech Republic: Elsevier Ltd; 2016, 292–299.

[45] Tay Y. W., Panda B., Paul S. C., Mohamed N. A. N., Ming Jen T., Kah Fai L. 3D printing trends in building and construction industry: A review. *Virtual and Physical Prototyping*. 2017, 12(3), 261–276.

[46] Montemurro M., Catapano A., Doroszewski D. A multi-scale approach for the simultaneous shape and material optimisation of sandwich panels with cellular core. *Composite B: Engineering*. 2016, 91, 458–472.

[47] Javaid M., Haleem A. Additive manufacturing applications in orthopaedics: A review. *Journal of Clinical Orthopaedics and Trauma*. 2018, 9(3), 202–206.

[48] Bhoi S., Prasad A., Kumar A., Sarkar R. B., Mahto B., Meena C. S., Pandey C. Experimental study to evaluate the wear performance of UHMWPE and XLPE material for orthopedics application. *Bioengineering*. 2022, 9, 676. 10.3390/bioengineering9110676.

[49] Kumar A., Gangwar A. K. S., Kumar A., Meena C. S., Singh V. P., Dutt N., Prasad A., Gori, Y. Biomedical study of femur bone fracture and healing. In *Advanced Materials for Biomedical Applications*; Kumar A., Gori Y., Kumar A., Meena C. S., Dutt N., Eds. Boca Raton, FL, USA: CRC Press. 2022; Chapter 14; 235–250.

[50] Datta A., Kumar A., Kumar A., Kumar A., Singh V. P. Advanced materials in biological implants and surgical tools. In *Advanced Materials for Biomedical Applications*; Kumar A., Gori Y., Kumar A., Meena C. S., Dutt N., Eds. Boca Raton, FL, USA: CRC Press. 2022; Chapter 2; 21–43.

[51] Prasad A., Chakraborty G., Kumar A. Bio-based environmentally benign polymeric resorbable materials for orthopedic fixation applications. In *Advanced Materials for Biomedical Applications*; Kumar A., Gori Y., Kumar A., Meena C. S., Dutt N., Eds. Boca Raton, FL, USA: CRC Press. 2022; Chapter 15; 251–266.

[52] Kumar A., Datta A., Kumar A. Recent advancements and future trends in next-generation materials for biomedical applications. In *Advanced Materials for Biomedical Applications*; Kumar A., Gori Y., Kumar A., Meena C. S., Dutt N., Eds. Boca Raton, FL, USA: CRC Press. 2022; Chapter 1; 1–19.

[53] Kumar A., Gori Y., Rana S., Sharma N. K., Yadav B. FEA of humerus bone fracture and healing. *Advanced Materials for Biomechanical Applications* (1st ed.). CRC Press. 2022, 10.1201/9781003286806.

[54] Gangwar A. K. S., Rao P. S., Kumar A. Bio-Mechanical Design and Analysis of Femur Bone. *Materials Today: Proceedings*. 2021, 44(Part 1), 2179–2187, ISSN 2214-7853 10.1016/j.matpr.2020.12.282.

[55] Bhoi S., Kumar A., Prasad A., Meena C. S., Sarkar R. B., Mahto B., Ghosh A. Performance evaluation of different coating materials in delamination for micro-milling applications on high-speed steel substrate. *Micromachines*. 2022, 13, 1277. 10.3390/mi13081277.

[56] Patil P. P., Sharma S. C., Jaiswal H., Kumar A. Modeling influence of tube material on vibration based EMMFS using ANFIS. *Procedia Material Science*. 2014, 6, 1097–1103.

[57] Zhang X., Liou F. Introduction to additive manufacturing. *Additive Manufacturing*. Elsevier. 2021, 1–31.

[58] Goulas A., Engstrøm D. S., Friel R. J. Additive manufacturing using space resources. *Additive Manufacturing*. Elsevier. 2021, 661–683.

[59] Vranić A., et al. Advantages and drawbacks of additive manufacturing. *IMK-14–Research & Development in Heavy Machinery*. 2017.

Chapter 4

Laser-Based Additive Manufacturing

Prameet Vats, Kishor Kumar Gajrani, and Avinash Kumar

CONTENTS

4.1   Introduction 67
4.2   Laser in Additive Manufacturing 68
       4.2.1   Classification of Laser-Based Additive Manufacturing 68
               4.2.1.1   Polymer-Based Additive Manufacturing 69
               4.2.1.2   Metal-Based Additive Manufacturing 70
       4.2.2   Design for Additive Manufacturing Processes 72
       4.2.3   Post-processing of Additive Manufactured Components 73
4.3   Defects in Laser-Based Additive Manufacturing Processes 75
4.4   Application of Laser-Based Additive Manufacturing Processes 77
       4.4.1   Medical and Dental Applications 77
       4.4.2   Automobile Application 77
       4.4.3   Aerospace and Military Application 78
4.5   Comparison of Laser-Based Additive Manufacturing 78
4.6   Current Issues, Challenges, and Future Scope 80
Acknowledgement 81
References 81

4.1 INTRODUCTION

Since the first development of additive manufacturing (AM) in the early 1980s, this technology has become a revolution in the manufacturing of complex products [1]. It has some key advantages such as design flexibility, cost and speed, which make it the centre of attraction for manufacturing industries [1,2]. In AM, components are fabricated layer by layer. The material is progressively added in each layer and part is produced; thus, the process is collectively termed as AM [2]. Nowadays, AM is widely adopted in various industries such as aviation, aerospace, automobile, biomedical, tool making, etc. [3].

AM is mainly classified into processes such as fuse deposition modelling, binder jetting, metal extrusion, material jetting, powder bed fusion, vat

DOI: 10.1201/9781003402398-4

photo-polymerisation, powder bed fusion, sheet lamination, etc. Each of them has a different external power source like laser beam, electron beam, and plasma arc. In vat polymerisation, powder bed fusion, and directed energy deposition laser is used as power source. In the last couple of years, laser additive technology has made some promising strides that have a lot of potential for industrial uses [1,3]. Therefore, nowadays industries are more focused on laser-based AM.

4.2 LASER IN ADDITIVE MANUFACTURING

In the year 1968 at the United States, first time a precursor AM process used a laser source. It was proposed to analyse the geometry of an existent thing without disturbing it and to replicate the geometry of the object using AM by employing crossed layers in a volume printing process. After a while, in 1972, a French patent application with the silent feature of "directed energy deposition" was submitted [1].

A laser is typically consisting of a gain medium, an optical resonator, and a pumping energy source. The pumping source provides energy from the outside, which is used by the gain medium to make the light beam stronger. It mainly exists inside the optical resonator. The list of few lasers which are mainly used in AM are CO_2 laser, Nd:YAG, laser, excimer laser, and Yb-fibre laser [4]. The laser-based AM processes generally use metal and polymers as the workpiece materials. The laser beam can immediately shift a huge amount of energy into a micro-scale focused zone to solidify or cure the working material. Because of the high energy, appropriate melting of the material can occur, resulting in a strong link between the layers [3].

4.2.1 Classification of Laser-Based Additive Manufacturing

Classification of laser-based AM was first suggested in ISO 17296. As per ISO 17296, it first distinguishes between the AM process on the basis of input feed materials such as ceramics, metals, biological materials, and polymers [3]. Laser-based AM processes have similar manufacturing principles throughout each process, but they can be differentiated on the basis of their procedure, input parameters, and input materials. The purpose behind using a laser in AM is to provide the heat source for the melting and combining of materials or in some cases like vat polymerisation, it is used to supply the light beam of a certain wavelength to the initial chemical curing reaction. Materials used in laser-based AM can be powder (metals, polymers, and ceramics), liquid (resin), or solid (plastics, paper, and metals) [5]. Most of the work performed in AM is on polymer and metal-based materials. The overview of laser-based AM for metal and polymer materials is discussed in the following subsections.

4.2.1.1 Polymer-Based Additive Manufacturing

Within the last decade, there have been some categories of polymer-based AM using a laser beam as a power source such as laminated object manufacturing (LOM) and stereolithography (SLA). Stereolithography (SLA) is a type of AM that builds parts by moving a vat liquid photo-polymer resin over a platform and by using laser beam that is controlled by a computer system. The appropriate wavelength of an electromagnetic wave is used to cure and solidifies when it comes into contact with photopolymers [6]. Therefore, most lasers are low-powered. The construction platform is then progressively lowered by an amount equivalent to the thickness of the layer. Then, a fresh resin film is applied on top of the previously hardened layer. This technique is performed layer by layer until the component is finished.

SLA has many advantages, such as a smooth surface finish and high dimensional accuracy. Thus, it has many applications such as prototyping and printing of non-structural parts with excellent surface finish and high dimension accuracy. The resins used in SLA mainly include epoxy macromeres and low molecular weight polyacrylate. Some non-reactive diluents, like water or methyl pyrrolidone, can be added to resins to make them less viscose [1,5].

On the other hand, SLAs have some limitations such as they need support structures, are expensive, can't work with metals, and have a limited supply of resins. However, for the printing of micro-structural parts, mostly micro-stereolithography is used. The schematic diagram for SLA is shown in Figure 4.1 [5].

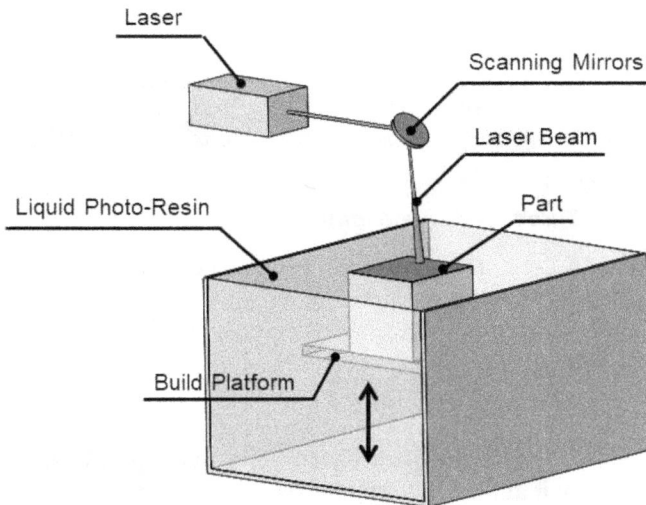

Figure 4.1 Schematic diagram of stereolithography (SLA) [5].

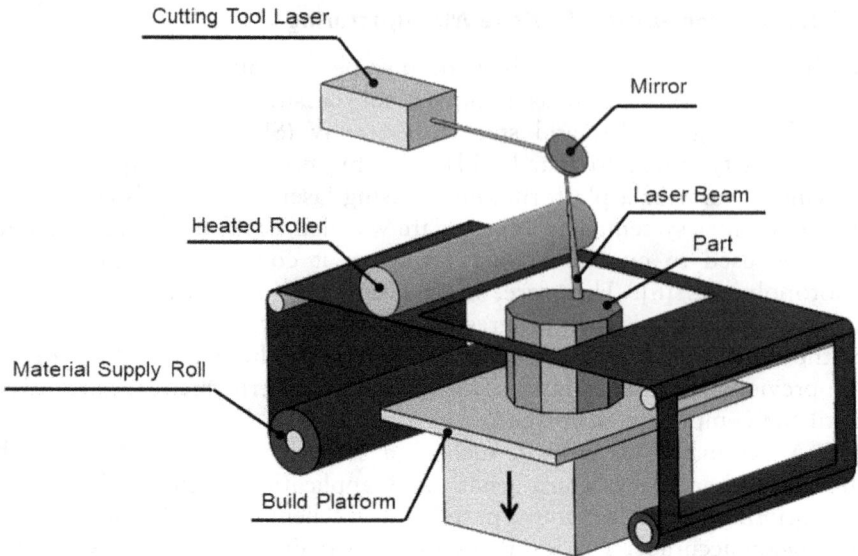

Figure 4.2 Schematic diagram of laminated object manufacturing [5].

In LOM, heated rollers are used to put a layer of adhesive material on the platform. The required cross-section is cut with a computer-controlled CO_2 laser, and a new layer has been added to the platform. The process is done over and over until the part is done [3]. The schematic diagram to show the procedure of LOM is shown in Figure 4.2 [5].

Material that is used in the LOM process is sheet metals, composites, synthetic materials, fabrics, and also plastics. LOM has some demerits, which include waste of material, waste of energy and money in post material removal process. The main problem is that the laser beam is hard to control, which can cause the dimensions to change and increase power consumption [7]. Ultrasonic consolidation and resistance welding have been added to this process to make it possible to print metal parts.

4.2.1.2 Metal-Based Additive Manufacturing

Metal-based AM is the most common process because of its ability to produce highly dense metallic part with complex geometry and better mechanical properties as compared to conventionally fabricated parts [8]. One of the methods to fabricate complex metal-based parts is laser-based melting (LBM). LBM machines need a high-intensity solid laser source, a building base, a powder-feeding container, a powder-depositing roller or scraping, and a beam-deflection mechanism. The first step in making a part is to make a 3D CAD model of it and turn it into the standard file format (.stl) for the AM manufacturing system [1]. Another important step for powder materials is to heat them before the laser strikes them. This improves the powder's laser

Types of material	**Metallic**			
State of fusion	Melted state		Solid + Melted state	Solid state
Material feedstock	Filament / wire material	Powder material		Sheet material
Material distribution	Deposition nozzle	Powder material		Sheet stock
Basic AM principle	Selective deposition of material to substrate	Selective fusion of material in powder bed		Fusion stocked sheet
Source of fusion	Electron beam / Laser	Electron beam / Laser		Ultrasound
Process Category	Directed energy deposition	Powder bed fusion		Sheet lamination

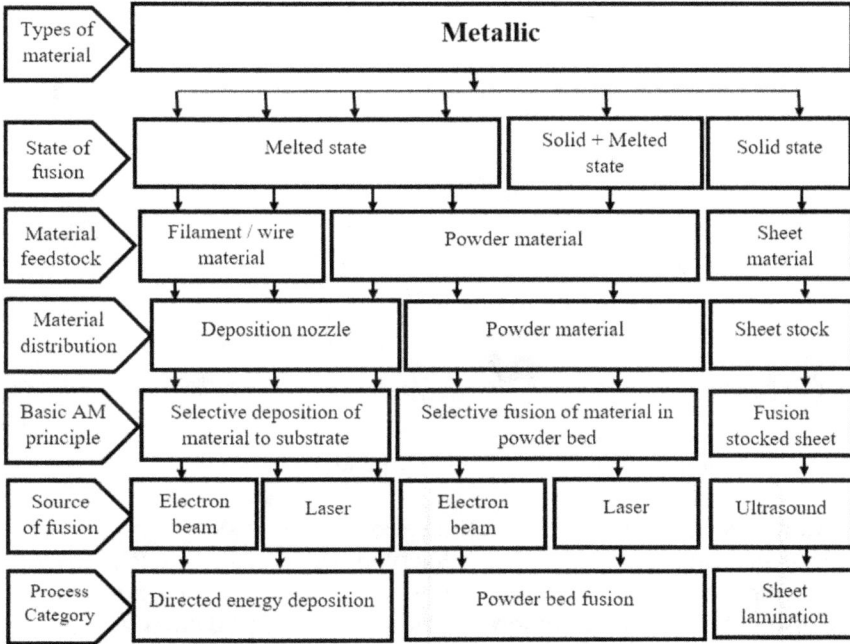

Figure 4.3 Classification of metal-based additive manufacturing.

absorption, reduces temperature differences, and improves wetting qualities. In the process, the thermal fusible power is used and preheated to temperatures lower than the sintering point. Figure 4.3 illustrates the classification of metal-based AM processes.

In laser sintering (LS), the power is moderately melted or heated to the sintering temperature. The components used in LS are similar to the LBM machines. In this process, heat fusible powder is employed, and pre heat is delivered at temperatures below the sintering point prior to scanning [6–8]. The scanning of the beam offers additional energy for particle partial melting or sintering. LS is mainly classified into two parts: (i) direct and (ii) indirect methods. In the indirect method, powder is mainly mixed with the binders to improve the sintering process, which helps to improve the density and strength of the material. Liquid phase sintering occurs when the binders are in a liquid condition during sintering. Binders are not necessary in the direct technique since the laser scanning source is enough to bind the particles.

In the case of laser metal deposition (LMD), a nozzle is used to provide powder. LMD is also known as direct metal deposition (DMD), laser deposition welding (LDW), laser-engineered net shaping (LENT), laser cladding, and direct energy deposition (DED). LDM uses a laser beam to focus on the workpiece, thereby producing the melt pool, whereas, weld trackers are inserted simultaneously with metallic powder in the melt pool to produce a single layer. Several layers are stacked on top of each other to

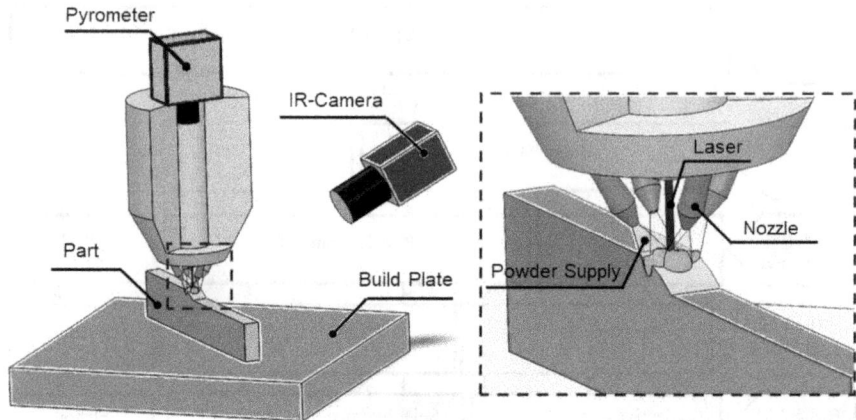

Figure 4.4 Schematic diagram of direct metal deposition [5].

create the shape [5]. The layer thickness is mainly dependent upon the process parameters (laser power, velocity, powder feed rate, etc.). In LMD, 0.5–3 mm laser beam diameters are used to confine the heat-affected zone and also to help in the reduction of residual stresses. LMD may also be used to repair worn-out high-value components, as well as for coatings to protect the product from wear and corrosion [8]. The schematic diagram of DMD is shown in Figure 4.4.

In a wire-based deposition system, wire is used to build the part. It is mainly used to build large components with moderate complexity. Wire-based AM process has various advantages over powder-based methods such as a higher rate of deposition, clean working environment, efficient consumption of resources, and more economical than powder [9].

4.2.2 Design for Additive Manufacturing Processes

AM is a totally different process as compared to the conventional manufacturing process. AM has their own design constraints and it also provides a new degree of freedom for design. The design of AM requires the development of new adapted design rules which are basically summarised on the basis of geometrical standards [1]. The basic standard components are classified into three categories.

- Basic elements: Basic elements serve as the foundation for both aggregated structures and element transitions. These are basic, mathematically defined geometrical forms that emerge when a profile is moved along a guide curve. As a result, three types of basic components may occur: double-curved, simple-curved, or non-curved elements.
- Element transition: For creating the technical pieces, predefined basic elements must be coupled with each other. As a result, element

transitions occur. Due to the layer-by-layer fabrication, two kinds of element transition may be distinguished: Transitions between securely bound element transitions between non-bonded components. The geometry of both transition types may vary significantly based on the integrated basic element types and the concreteness of the elements.

- Aggregated structures: Aggregation structures define the spatial arrangements of fundamental constituents and their transitions. Equal aggregated structures may be formed using various combinations of basic components, similar to element transitions.

In order to determine suitable ranges, standard elements can be made by varying parameters. Also, the quality of the product that are made can be checked and linked to the parameters. In the end, the rules for designs are made based on the results [10].

Parts produced by the AM show the poor surface quality. There are several aspects that might influence the surface quality of a product during the AM process, including process parameters, design guidelines, and so on. However, experimentally determined designed rules may aid to enhance the surface quality of the product. There has been some experimental research has been conducted in order to generate design principles for AM of metals and polymers based on geometrical elements. As per the research, the cross-sectional area of the building parts should remain the same or get smaller so that thermal energy is distributed uniformly. When the cross-sectional areas of the layers are expanded in comparison to the initial cross-sections, discoloration develops because of excess heat energy. It was also mentioned that rounded and blunted edges should be preferred over sharp edges for uniform distribution of thermal energy.

Parts having enclosed voids must be avoided because the surplus material is removed after the part is fully constructed. The substance cannot be removed if the void is enclosed. In order to reduce cost, mass and to improve the strength and energy absorption characteristics of AM parts, a lattice structure can be employed. Lattice structures have a combination of high strength, low weight, and tremendous energy absorption, which make them best suited for high-value medical applications. and aeronautical components [1,10].

4.2.3 Post-processing of Additive Manufactured Components

Post-processing techniques are often performed to improve the mechanical properties and surface quality of the product. Many post-processing technologies have been developed, such as thermal post-processing techniques for releasing thermally generated residual stress, laser peening for reducing micro-defects as well as improving surface quality, and machining for improving product surface quality [6]. Some of the most frequent post-processing procedures include thermal/heat post-processing, machining, laser polishing, laser peening, and abrasive finishing.

Laser peening is a technique that plastically compresses material perpendicular to the surface to induce lateral expansion [9]. When laser peening is used on thick or restricted objects, the capacity to withstand transversal strain results in the development of local compressive stresses. In the case of thin parts, laser peening causes shape and change in strain. All deformation-based post-processing techniques include lateral expansion and plastic compression. Laser peening is commonly used to increase the fatigue life and surface characteristics of materials [11]. Currently, its major part of the application is in compression blades, jet engine fans, aircraft structures and nuclear-spent fuel storage tanks and also in thick sections of the aircraft fenders to make sure that the aerodynamic models are correct. The schematic diagram of laser peening is shown in Figure 4.5 [9].

In order to improve residual stresses, reduce cracking, and homogenising the microstructure of AM components, thermal post-processing is used. Hot isostatic pressing (HIP), solution heat treatment (SHT), and T6-heat treatment are all thermal processing methods (T6-HT). Numerous investigations on the mechanical properties and microstructure of AM parts have been conducted [12]. In a typical thermo mechanical process, high temperature and pressure are used in conjunction to remove pores and to improve the density of the finished pieces. The typical working temperature for this technique is between 1000 and 2000°C. HIP aids in the production of components with high density, excellent uniformity, and superior performance. HIP also optimises the mechanical properties of LBM and electron beam melting (EBM) components. Due to the reduction in unmelted material and porosity, as well as the coarsening of the microstructure, the HIP treatment may result in components with improved mechanical qualities [13].

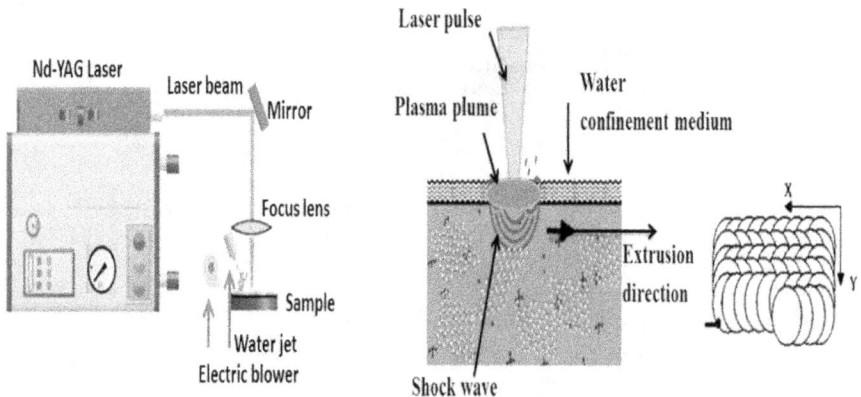

Figure 4.5 Schematic diagram of laser peening [9].

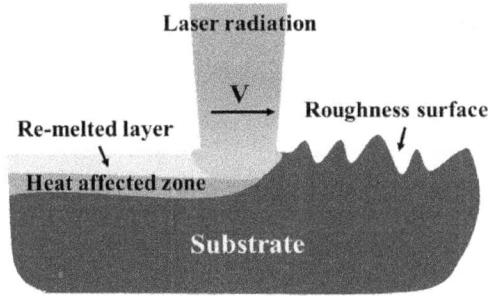

Figure 4.6 Schematic of laser polishing [12].

Laser polishing is seen as a potential technology for reducing the surface roughness of AM components. When the energy source shines on the material's surface during laser polishing, the morphological peaks might soon approach the melting point as shown in Figure 4.6. Once a molten pool forms, the liquid moves back to the same level because of gravity and surface tension [12]. The temperature of the heat-affected zone (HAZ) rapidly declines after the laser beam stops scanning the surface. This causes the molten pool to solidify, which makes the surface smoother. Laser polishing is a machine-controlled process that changes the shape of the surface by re-melting it. This doesn't change or affect the bulk properties of the material [13].

Machining and abrasive finishing are two popular methods used by manufacturers across numerous sectors to enhance the form and quality of working components. They are well-known and simple to use and are often employed as post-processing procedures for AM components [9]. Most of the time, grinding and milling are used to improve the quality of the fabricated part surface. Milling may significantly reduce the roughness of the workpiece surface and feed rate. The high cutting speed helps to improve the surface quality [13]. The magnetic abrasive finishing (MAF) process is used to polish the surface of an AMed object. Surface defects like unmelted atoms and balling are removed, and the surface quality may be enhanced by up to 75%.

4.3 DEFECTS IN LASER-BASED ADDITIVE MANUFACTURING PROCESSES

Stainless steel, nickel alloys, aluminium, and titanium alloys, are the most frequently used materials in AM process. AM defects are most often caused by the good thermal conductivity and reflectivity of alloys. The major defects in AM are pores, anisotropy, residual stress, cracks, thermal stresses, surface roughness, laser spatting, and shape distortion. Various parameters are responsible for defects in components fabricated using AM process.

One of the major drawbacks of AM components is the formation of voids among successive layers. This kind of issue comes as a result of inappropriate heat deposition on the material surface, which reduces bonding between layers, resulting in poor mechanical performance. Indeed, the quantity of porosity caused by void formation is generally determined by the kind of AM method employed and the material used. As a result, to reduce the influence of the development of a void between the following layers, the impact of nozzle geometry and laser intensity on the development of voids between succeeding layers must be minimised. According to the findings of their investigation, rectangular nozzles are more efficient than cylindrical ones, and an increase in laser power with a low speed might be a better combination for effective heating [14].

Pores are small size gas holes approximately spherical in shape. The formation of pores is a major flaw in the AMed components. Porosity is categorised into two types: (i) metallurgical pores formed by adsorption of surrounding gases, and (ii) parameter-based pores caused principally by successive dilation cycles resulting in local failure. As a consequence, macro-porosities are more harmful than micro-porosities since they do not have a spherical form and may induce fractures. Micro-porosity may become crucial after the heat treatment procedure since it has a tendency to combine during such operations and generate macro-porosities, which, as previously noted, are the source of fractures [15]. Porosity is caused due to several factors, including process parameters (laser power, hatch size, scan speed, etc.), the presence of impurities in the base material, the restricted absorption of laser energy by materials, and the atmospheric condition of the combustion chamber.

Crack formation is another type of defect encountered during the AM process. The absence of molten material supply to the inter-dendritic gaps causes the formation of voids during the solidification of the material. A vast range of solidification may be caused by this effect. To prevent the production of voids, alloy compositions are altered by adding certain elements. A procedure of trial and error should be used to locate the element. The addition of silicon to the aluminium alloy improves the eutectic phase and lowers the solidification range and melting temperature, hence reducing the risk of hot cracking in the alloy [16].

Poor surface finish is the major issue in AMed fabricated components. The surface finish is mostly affected by the process parameters such as laser power, hatch size, scan speed, material deposition speed, etc. The deterministic link between operating circumstances and surface quality may be identified to tackle this challenge. The laser parameter settings have a direct impact on the stability of the melt pool, which results in surface uniformity. Scan speed is also important; a slow laser scan speed increases surface roughness [15,16]. A high metal deposition rate causes poor surface finish during part printing. Parameters like print angle, powder size, and hatch spacing also affect the surface roughness in AMed components.

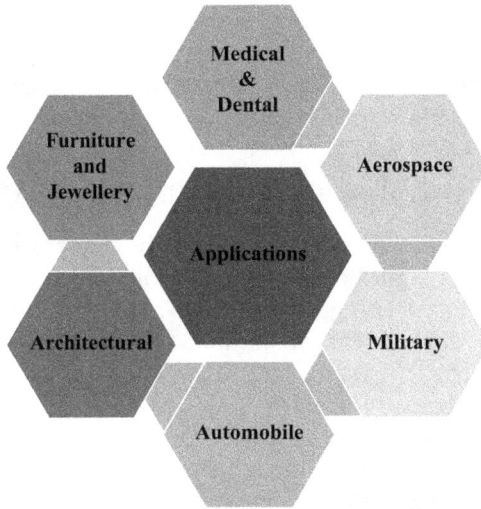

Figure 4.7 Application of additive manufacturing.

4.4 APPLICATION OF LASER-BASED ADDITIVE MANUFACTURING PROCESSES

AM has applications in the aerospace, military, automobile, medical, fashion, dental sectors, biotechnology, food industries, eyewear, spatial analysis, education, etc. Figure 4.7 shows the various applications of AM.

4.4.1 Medical and Dental Applications

AM in the medical and dental field can be used to create testing and medical devices, tissues, organs, implants etc. Nowadays, AM is being utilised to construct surgical instruments and patient-specific organs. Anatomical templates are among the most widely used AM technologies in the medical field. The AM anatomy framework assists clinicians in making better treatment decisions and more accurately planning surgeries. Scientists will employ additive printing technology to mass-produce functional human bodies, even procedure organs, and replace joint cartilage, according to biomedical corporations [17].

4.4.2 Automobile Application

AM is increasingly being used in all aspects of automobile manufacturing. In addition to AM, the technique is utilised to produce tools and finished components. AM enables automotive designers to swiftly manufacture physical items or assemble prototypes, which may vary from a small interior part to a dashboard or even a scale model of a complete automobile

[18]. When compared to more conventional ways, AM in the automobile design industry uses fewer resources and creates less waste. Also, it saves time and energy at different phases of manufacturing and reduces overall production costs. [19].

4.4.3 Aerospace and Military Application

Aerospace components that value aesthetics above performance, like power wheels, door handles, light housings, and whole interior dashboard designs are typically produced using AM. Metal-based AM enables the rapid manufacture and deployment of complex military components such as bunkers, helmets, etc. The major role of AM is observed in the production of obsolete components, which are no longer supplied by original manufacturers [20]. For the Indian military, AM services enable on-site customisation of replacement components to solve this issue. AMed aerospace and military equipment components are gaining popularity. One of the most appealing aspect of AM is its low operating costs. Faster turnaround times and cheaper costs may be achieved by using AM to modify military equipment. One of the most compelling aspects of AM for defence is its ability to reduce waste. For example, titanium is a valuable metal that can be efficiently and effectively utilised in 3D printing without generating a lot of waste [17,20].

4.5 COMPARISON OF LASER-BASED ADDITIVE MANUFACTURING

Over the last decade, the AM process has evolved rapidly. Using less energy and resources in production means a reduction in the quantity of raw materials needed in the supply chain, which in turn frees up resources for more environmentally friendly practices [1]. Furthermore, AM has numerous advantages over conventional manufacturing process such as design flexibility, need for assemblage, dimensional accuracy, time, cost of geometric complexity, and production cost efficiency.

AM also has several classifications and each one has different power sources such as electron beam, laser beam, thermal heat source, UV, etc. Among all the power sources most of the AM techniques use laser as a power source. As a result, laser power is preferred above other power sources in the AM industry since it is the most efficient. High-precision and high-volume production of a broad variety of materials may be achieved with the use of a laser beam, which can rapidly transfer a huge quantity of energy into a micro-scale focused zone to cure and solidify materials in the air [21]. Table 4.1 shows the comparison between different AM technologies on the basis of power sources, processing materials with its pros and cons.

Table 4.1 Comparison between the different additive manufacturing process

Technologies	Power source	Materials	Pros and cons
Fused deposition modelling	Thermal	Thermoplastics, metal pastes, ceramic slurries	• Low-cost extrusion machine • Printing on many materials • Poor surface finish
Selective laser melting/ electron beam melting	High-powered laser beam, electron beam	Atomised metal powder (stainless steel, cobalt, chromium, titanium), polyamides	• High accuracy • Full dense component • Excellent specific strength and stiffness • Recycling and power handling • Structural support and anchorage
Stereolithography	Ultraviolet	Photopolymer, ceramics	• Fast construction • Good component resolution • Overcuring, scanned line forms • Expensive supply-materials cost
Polyjet/ Inkjet Printing	Thermal energy/ photocuring	Photopolymer, wax	• Printing on several materials, • High surface polish, • Low strength material
Indirect inkjet printing	Thermal energy	Polymer powder, ceramic powder, metal powder	• Full-colour object planning. • Wide material selections • High porosity on finished part
Laminated object manufacturing	Laser beam	Ceramic tape, plastic film, metallic sheet	• Excellent surface polish • Low melting, machining and process cost
Laser-engineered net shaping, electronic beam welding	Laser beam	Molten metal powder	• Repair of damaged/ worn components • Printing on functionally graded material • Require post-processing machine

4.6 CURRENT ISSUES, CHALLENGES, AND FUTURE SCOPE

Certainly, there are several benefits to adopting AM, including as creative flexibility, the capacity to fabricate complex patterns/structures, easy to use, and product personalisation. However, AM process has not yet evolved enough to be applied in all practical applications. There are limitations and obstacles that need more research and technological advancement. Some of the challenges that must be addressed include size limitations, anisotropic mechanical qualities, fabrication of overhang surfaces, low manufacturing efficiency, higher prices for more volume, and warping and poor precision [21]. Stringing and pillowing are also issues that need to be addressed as well as gaps in the top layers, under-extrusion, layer misalignment, and elephant feet.

The occurrence of staircase effect or stacking mistake in manufactured parts is one of the most difficult issues in the AM process. This kind of mistake is minor for internal manufactured surfaces, but it has a considerable effect on the quality of exterior surfaces. Although several technologies (post-processing) such as sand sintering may be used to decrease or eliminate this defect, they also increase the overall time and expense of the operation [22].

Another issue with AM is the presence of anisotropy in the mechanical characteristic and microstructure. AM processes create component layer by layer by curing the photo resin, melting filaments, or heating the powder bed, resulting in a temperature gradient. AM components often exhibit different mechanical properties and microstructures in the build path and in other directions.

In the future aspects, the key consumers of AM are not likely to alter considerably in the coming years. The transportation (aerospace and automobile) and medical industries, in particular, are expected to continue to be key consumers. Unless and until a broader variety of businesses are incentivised to adopt AM in the future, the forecast that AM will transform the way industries interact and global supply chains work on a worldwide scale will be incorrect. To be successful, a company's organisation, and indeed its whole culture, must adjust to the transition from conventional to AM [22]. According to Gilbert's research, AM has trouble entering high-risk sectors with significant competition. Potential obstacles include routine rigidity (failure to change organisational procedures that employ those resources) and resource rigidity (the inability to alter resource investment patterns) [23]. The two more major aspects or possibilities that might encourage further use of the technology: (a) sustainability in terms of environment and (b) opportunities for integration

The environmental effects of manufacturing are becoming more important, and the future is expected to witness rising demands on material and energy usage, as well as greenhouse gas emissions. This could be a driving force for AM in the future since studies show that AM is more efficient in terms of using raw materials and water, making less pollution, and needing less landfill space in most situations where it is economically feasible to use AM [24].

A worldwide architectural shift in marketing or the end of conventional manufacturing techniques industries are moving towards AM. AM gives a wide variety of parts that may be produced using this technology [25]. This hints at a future in which production is centralised to some extent and combines forming, subtractive, and additive processes.

As a result, the success factor "Combination of Manufacturing Technologies" indicated for all industrial sectors seems to be very important, with the manufacturing technologies presented going beyond simply additive. In some places, the integration of subtractive and additive technologies already occurs. As an example, microfabrics electrochemical fabrication method uses planarisation and multilayer electrodeposition of at least two metals, as well as a weld deposition system that can be adapted to any CNC machining centre [23,26]. However, it is not a well-developed technology at the moment. In the future industries will prefer the integrated manufacturing system that includes both the additive as well as subtractive methods.

ACKNOWLEDGEMENT

This work was supported by the Institute Seed Grant from the Indian Institute of Information Technology, Design and Manufacturing, Kancheepuram, India (No. IIITDM/ISG/2022/ME/02). Authors are also thankful to the copyright holders of various figures for providing copyright to reuse figures in this work.

REFERENCES

[1] Schmidt, M., Merklein, M., Bourell, D., Dimitrov, D., Hausotte, T., Wegener, K., Overmeyer, L., Vollertsen, F., Levy, G. N. (2017). Laser-based additive manufacturing in industry and academia. *CIRP Annals-Manufacturing Technology*, 66(2), 561–583.

[2] Sahasrabudhe, H., Bose, S., Bandyopadhyay, A. (2018). Laser-based additive manufacturing processes. *Advances in Laser Materials Processing* (pp. 507–539). Woodhead Publishing.

[3] Kumar, S., Pityana, S. (2011). Laser-based additive manufacturing of metals. *Advanced Materials Research* (Vol. 227, pp. 92–95). Trans Tech Publications Ltd.

[4] Lee, H., Lim, C. H. J., Low, M. J., Tham, N., Murukeshan, V. M., Kim, Y. J. (2017). Lasers in additive manufacturing: A review. *International Journal of Precision Engineering and Manufacturing – Green Technology*, 4(3), 307–322.

[5] Razavykia, A., Brusa, E., Delprete, C., Yavari, R. (2020). An overview of additive manufacturing technologies—A review to technical synthesis in numerical study of selective laser melting. *Materials*, 13(17), 3895.

[6] Mahmood, M. A., Chioibasu, D., Ur Rehman, A., Mihai, S., Popescu, A. C. (2022). Post-processing techniques to enhance the quality of metallic parts produced by additive manufacturing. *Metals, 12*(1), 77.

[7] Tofail, S. A., Koumoulos, E. P., Bandyopadhyay, A., Bose, S., O'Donoghue, L., Charitidis, C. (2018). Additive manufacturing: Scientific and technological challenges, market uptake and opportunities. *Materials Today, 21*(1), 22–37.

[8] Tapia, G., Elwany, A. (2014). A review on process monitoring and control in metal-based additive manufacturing. *Journal of Manufacturing Science and Engineering, 136*(6), 060801.

[9] Prabhakaran, S., Kumar, H. P., Kalainathan, S., Vasudevan, V. K., Shukla, P., Lin, D. (2019). Laser shock peening modified surface texturing, microstructure and mechanical properties of graphene dispersion strengthened aluminium nanocomposites. *Surfaces and Interfaces, 14*, 127–137.

[10] Adam, G. A., Zimmer, D. (2014). Design for additive manufacturing—Element transitions and aggregated structures. *CIRP Journal of Manufacturing Science and Technology, 7*(1), 20–28.

[11] Thompson, M. K., Moroni, G., Vaneker, T., Fadel, G., Campbell, R. I., Gibson, I., Martina, F. (2016). Design for additive manufacturing: Trends, opportunities, considerations, and constraints. *CIRP Annals- Manufacturing Technology, 65*(2), 737–760.

[12] Zhou, J., Liao, C., Shen, H., Ding, X. (2019). Surface and property characterization of laser polished Ti6Al4V. *Surface and Coatings Technology, 380*, 125016.

[13] Kumbhar, N. N., Mulay, A. V. (2018). Post processing methods used to improve surface finish of products which are manufactured by additive manufacturing technologies: A review. *Journal of The Institution of Engineers (India): Series C, 99*(4), 481–487.

[14] Khamidullin, B. A., Tsivilskiy, I. V., Gorunov, A. I., Gilmutdinov, A. K. (2019). Modeling of the effect of powder parameters on laser cladding using coaxial nozzle. *Surface and Coatings Technology, 364*, 430–443.

[15] Brennan, M. C., Keist, J. S., Palmer, T. A. (2021). Defects in metal additive manufacturing processes. *Journal of Materials Engineering and Performance, 30*(7), 4808–4818.

[16] Fu, Y., Downey, A. R., Yuan, L., Zhang, T., Pratt, A., Balogun, Y. (2022). Machine learning algorithms for defect detection in metal laser-based additive manufacturing: a review. *Journal of Manufacturing Processes, 75*, 693–710.

[17] Mohanavel, V., Ali, K. A., Ranganathan, K., Jeffrey, J. A., Ravikumar, M. M., Rajkumar, S. (2021). The roles and applications of additive manufacturing in the aerospace and automobile sector. *Materials Today: Proceedings, 47*, 405–409.

[18] Harun, W. S. W., Kamariah, M. S. I. N., Muhamad, N., Ghani, S. A. C., Ahmad, F., Mohamed, Z. (2018). A review of powder additive manufacturing processes for metallic biomaterials. *Powder Technology, 327*, 128–151.

[19] Horn, T. J., & Harrysson, O. L. (2012). Overview of current additive manufacturing technologies and selected applications. *Science Progress, 95*(3), 255–282.

[20] Liu, R., Wang, Z., Sparks, T., Liou, F., Newkirk, J. (2017). Aerospace applications of laser additive manufacturing. *Laser Additive Manufacturing* (pp. 351–371). Woodhead Publishing.

[21] Yan, Z., Liu, W., Tang, Z., Liu, X., Zhang, N., Li, M., Zhang, H. (2018). Review on thermal analysis in laser-based additive manufacturing. *Optics & Laser Technology*, *106*, 427–441.

[22] Abdulhameed, O., Al-Ahmari, A., Ameen, W., Mian, S. H. (2019). Additive manufacturing: Challenges, trends, and applications. *Advances in Mechanical Engineering*, *11*(2), 1687814018822880.

[23] Pinkerton, A. J. (2016). Lasers in additive manufacturing. *Optics & Laser Technology*, *78*, 25–32.

[24] Gao, W., Zhang, Y., Ramanujan, D., Ramani, K., Chen, Y., Williams, C. B., Wang, C. C., Shin, Y. C., Zhang, S., Zavattieri, P. D. (2015). The status, challenges, and future of additive manufacturing in engineering. *Computer-Aided Design*, *69*, 65–89.

[25] Levy, G. N. (2010). The role and future of the laser technology in the additive manufacturing environment. *Physics Procedia*, *5*, 65–80.

[26] Crafer, R., Oakley, P. J. (1992). *Laser Processing in Manufacturing*. Springer Science & Business Media.

ABBREVIATION

AM	Additive manufacturing
AMed	Additive manufactured
CAD	Computed-aided design
DED	Direct energy deposition
DMD	Direct metal deposition
EBM	Electron beam melting
HAZ	Heat-affected zone
HIP	Hot isostatic pressing
LBM	Laser-based melting
LDW	Laser deposition welding
LENT	Laser-engineered net shaping
LMD	Laser metal deposition
LOM	Laminated object manufacturing
LS	Laser sintering
MAF	Magnetic abrasive finishing
SHT	Solution heat treatment
SLA	Stereolithography

Chapter 5

Prospects of AI and ML in Laser-Based Manufacturing Technologies

*Khaja Mohiddin Shaik, Yalavarthi Suresh Babu,
Sanjay Gandhi Gundabatini, and
Vedantham Ramachandran*

CONTENTS

5.1 Introduction 85
5.2 Role of ML in LBM 90
    5.2.1 DL-Based Predictive Visualization of Fiber
        Laser Machining 92
    5.2.2 Using ML, the Effects of Laser Energy Uncertainty on
        Temperature Changes in Directed Energy Deposition 94
    5.2.3 ML-Based Challenges 95
5.3 Role of AI in LBM 96
    5.3.1 AI Challenges 100
5.4 Conclusion 102
References 102

5.1 INTRODUCTION

The industrial revolution, which began in the 18th century with the discovery of the steam engine and continued with the development of information and communication technologies, has caused numerous paradigm shifts in the factory industry in the past. In order to prepare for enhancing their portability by eliminating ineffective operations to acquire a higher production, the world's manufacturing companies are currently concentrating on smart manufacturing domains. Moving on, mechanization and robotization came after industry 4.0 in the late 20th century. Robots, cloud computing, 3D printing, IoT, deep learning (DL), artificial intelligence (AI), and other significant advancements have all been used to increase productivity in a range of manufacturing businesses [1]. A high level of technological advancement in the field of sophisticated smart systems has paved the way for increased usage of cutting-edge technologies like machine learning (ML) algorithms in the manufacturing sector. Figure 5.1 illustrates the many stages of problem-solving by AI. Gathering the data and information required to pre-process for the following analysis is vital when

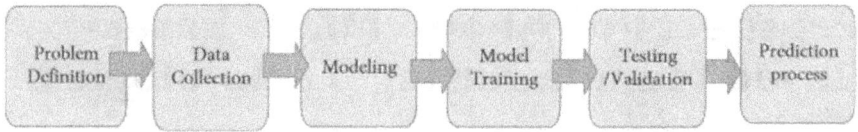

Figure 5.1 The procedures for applying artificial intelligence to solve problems.

choosing a ML method to answer properly stated challenges. To get the greatest outcomes, the appropriate model must be created and thoroughly trained using the necessary amount of data, in accordance with the information acquired. In order to acquire the best-improved results, it is also necessary to study and validate the obtained results.

John McCarthy [2] first used the term artificial intelligence (AI) to describe the intelligence exhibited by robots. The concept of AI was first proposed by the British mathematician and philosopher Scientist Alan Turing before McCarthy and computer. This all began with the straightforward inquiry, "Can robots think?" [3], in which he explained through an emulation game the rational process of breaking down the knowledge piece by piece to reach a wise conclusion. Prior to Turing, Pitts and McCulloch developed the foundation for neural networks by combining Turing's computational theory, Russell and Whitehead's propositional logic, and basic neurophysiology and functions. AI can be divided into symbolic or classical intelligence, which involves using knowledge and reasoning to solve issues, and computational intelligence, which involves using example data to solve problems and make decisions. According to the IEEE Computational Intelligence Society, artificial neural networks, fuzzy systems, and evolutionary programming are all included in computational AI (Figure 5.2).

Both symbolic and computational intelligence can be learned using various techniques, such as machine learning (ML), which uses both simulations and experiments. Automated reasoning [4] is one of the subfields in AI where computer programs are utilized to give computers the ability to fully

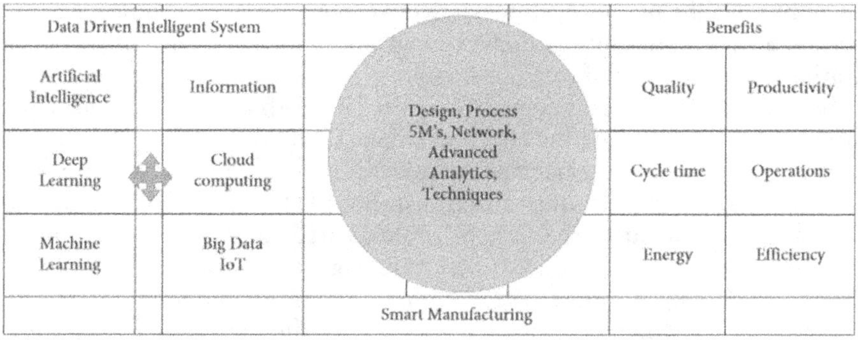

Figure 5.2 The components of smart manufacturing and its advantages.

or almost fully reason and act. The machine's logic or deductions may frequently need to be made in the face of uncertainty [5,6]. Making decisions in these circumstances is a probabilistic activity rather than a deterministic one, making complicated subjects like fuzzy logic and Bayesian statistics particularly useful in comprehending them.

There are an expanding variety of uses for lasers, including engraving, ablation, additive manufacturing, cutting, marking, cleaning, drilling, marking, and welding. Likewise, a wide variety of materials are processed using lasers, including a surprising number of ones like cotton, wood, leather, dough, and even egg shells. These materials include ceramics, metals, polymers, thin films, composites, glasses, and anodized surfaces. Almost every manufacturing sector uses laser materials processing, including those in the aerospace, automotive, batteries, electronics, medical, 3D printing, sensors, semiconductors, and solar industries. Similarly, size scales range from those used for welding ship steel all the way down to those used for processing materials close to the diffraction limit. In the decades since its inception, lasers have, to put it simply, revolutionized manufacturing. Due in part to the remarkable versatility with regard to characteristics such as pulse energy, pulse length, beam size, and wavelength, there are an astonishing variety of academic and industrial applications. The price of this flexibility, however, is the considerable effort and time required to discover the perfect number of criteria for each production process. For instance, if you need to modify five parameters by ten steps each, you would need to conduct 10^5 experimental trials. The norm is often the systematic collection of laser manufacturing data for all parameter categories in order to establish the ideal combination due to the sometimes very nonlinear relationships among parameters. This procedure, though, can take days or weeks and is unfocused and time-consuming, resulting in wasted money, effort, and time. Even after the ideal parameters have been determined, minute modifications in the manufacturing process, such as those to the laser's power or beam's shape, might cause the final product to fall short of the necessary standards, again at a cost in both time and money. Therefore, a set of modeling approaches is required for choosing the best settings, providing real-time monitoring, and correcting errors made during manufacturing. Finite element modeling is widely used in laser machining, which is a crucial enabling technology for many industrial manufacturing processes [7,8], involves the use of laser energy to selectively warm as well as soften materials before using a traditional cutting tool in laser-aided machining. Tagliaferri et al. work allowed for less cutting-tool wear and reduced cutting forces. Use the FEM to explain the laser heating process for laser power, speed, scanner, surface roughness, and focus distance in [9]. The FEM was used by Yang et al. [10] to forecast the depth and breadth of the heat-affected zone when laser warming a titanium alloy. In their study [4], Tian and Shin demonstrated how to use the FEM to model the development of defects like cracks in the shear zone as a result of laser-assisted machining of carbon nanotube ceramics. Singh et al. [5] reported using the FEM to predict temperature distribution, flow decreased stress from cutting forces and laser heating to optimize laser-assisted machining of steel.

A wide range of algorithms that continuously get better with use are included in the field of ML. Importantly, with such methods, the algorithms learn data attributes directly from the data rather than requiring an explicit programmatic definition of the relationship and the rules governing the data. Unsupervised, supervised, and reinforcement learning can be used to categorize the field of ML. The creation of a mathematical model for supervised learning entails mapping input data onto the corresponding output data so that predictions may be made. Unsupervised learning [6] is used to find clustering data points as well as structure within data. In order to maximize a reward, a software agent must follow the rules of reinforcement learning [11]. Although all three are significant topics for research, this chapter focuses on supervised learning because it has been the subject of most laser-machining studies so far. The field of DL resides inside ML, which is thought of as a subset of AI, as shown in Figure 5.3. In addition, the convolutional neural network (CNN) is a crucial component of DL. Another subtype, the conditional deep convolutional network (cGAN), is a neural network design made up of two CNNs that work together to train the network.

There are many different types of models in the field of ML [12], and the most well-known is briefly addressed next. An ANN is composed of a network of connected artificial neurons that are loosely modeled just after neurons in biological intelligence. As seen in Figure 5.4, an ANN typically consists of a hidden layer, an input layer, and an output layer. Each artificial neuron gets input from nearby neurons internally and processes the total

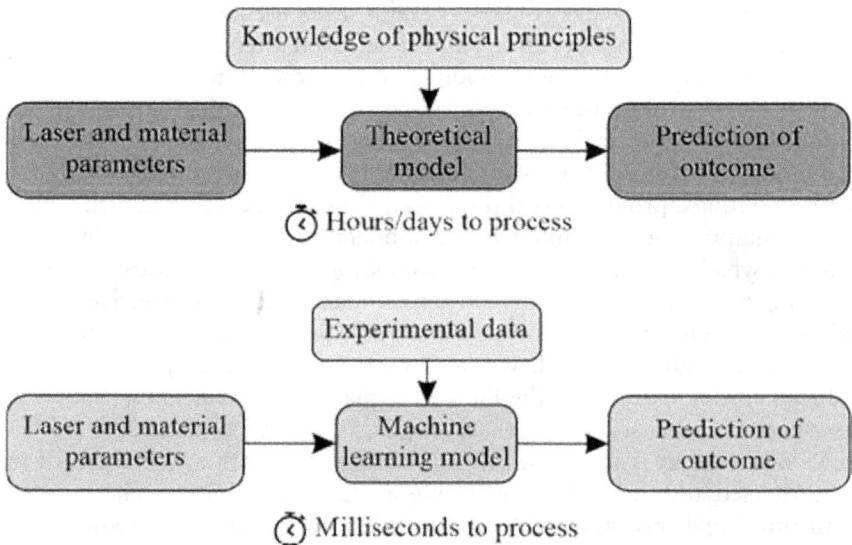

Figure 5.3 Time consumption for theoretical and ML models.

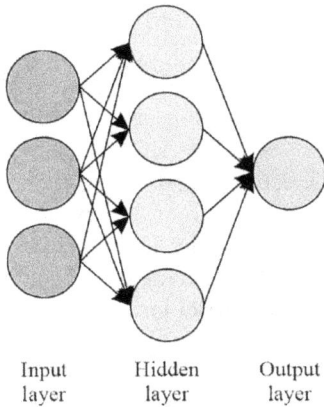

Input   Hidden   Output
layer   layer   layer   *Figure 5.4* Representation of neural networks.

amount of data using a nonlinear function before sending the resultant data onward through the network. Backpropagation, an algorithm, is often used in the learning process to automatically optimize the nonlinear functions at each neuron [13]. The network can therefore be utilized to convert input data into output data by acting as a transfer function. Decision trees [14] produce predictions about an item's value based on observations of the object by using a network of evaluations. Support vector machines enable the presentation of the data sample as points in space mapped in a way that makes a distinction among two or more subsets of the data sample obvious.

The same space can subsequently be used to map new data items, allowing a category prediction to be formed based on where they are in relation to the divider. A variety of statistical techniques, including both linear and nonlinear regression, are included in regression analysis which is used to evaluate the correlations among data inputs and related outputs. Using Bayesian networks, a probabilistic graphical linkage model can be created, allowing for the discovery of the interactions between variables as well as the prediction of a probability distribution of outcomes for each input. Natural selection influences the development of genetic algorithms, a type of search algorithm that increases the likelihood of discovering workable solutions to a specific problem. To produce new genotypes, these algorithms use algorithmic analogs of biological processes such as recombination and mutation effects.

In the domains of pattern recognition, data mining, and data science, where the differences between the three fields are mostly due to the implementation of their respective results, DL has demonstrated to be quite beneficial for addressing many challenging jobs. For instance, although jobs in data science are frequently industry-specific, those inside pattern recognition tends to be broader. Because of the advancement of heuristic training deep network techniques and the advancement of GPUs, deep models are now regarded as the industry standard for both academia and industry [15]. Particularly in the areas of computer vision, computational neuroscience, and medical image analysis, CNN architectures have delivered cutting-edge performance [16,17]. DNN architectures have also been used in NLP for a variety of tasks, including learning representation of words, machine translation [18], language comprehension [19], speech recognition, and sophisticated control systems [20]. The SGD method or second-order techniques, neural networks with a rising number of hidden layers, computing gradients that use the backpropagation algorithm, and the use of modular components are some recurrent themes in DL [21]. The usage of modules in DL has expedited the ability to develop task-specific modules that enhance the potent representational learning capabilities of DNNs. This is particularly evident in the field of computer vision, where CNN modules seek to reduce the total number of parameters to permit gradient flow during training and expand the region of interest of filters in upper layers.

5.2 ROLE OF ML IN LBM

Recently, ML has emerged as a significant alternative because it provides the ability to create an empirical approach directly from experimental results. In other words, ML makes it possible for laser machining to apply data-driven modeling. The fact that all empirical characteristics, including some that could be difficult to programmatically define, experiment noise, or like beam shape non-uniformity, are instantly incorporated into the modeling system is a key advantage of employing a data-driven modeling approach. The standard three phases for using neural networks for laser machining modeling are shown in Figure 5.5. The first stage involves

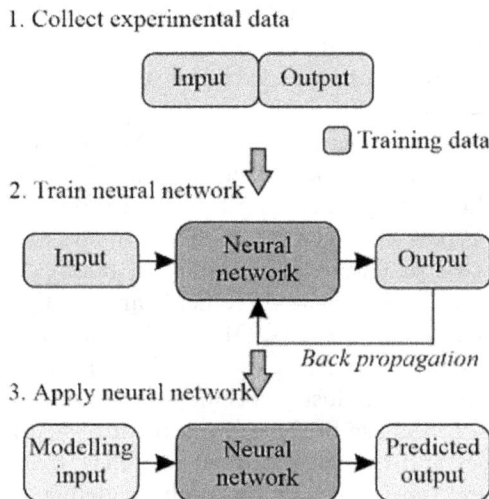

Figure 5.5 Idea of using ML in practical laser machining.

acquiring experiment information in the form of input data and matching data outputs. Consider employing sampling frequency and laser fluence as the inputs, with an output that quantifies the depth of the laser-cut features, as an example.

The neural network is trained in the second step, during which back-propagation is used to automatically alter the internal weights of the network. In order to maximize training progress, neural network accuracy is frequently assessed during training. The trained neural network is given a new set of input values in the third stage, and it predicts the associated output values. The neural network can be utilized as a "black box" modeling element that converts input data into the appropriate output data once it has been trained. Notably, the structure of the output and input data used for training directly influences the type of transition from input data to output data. In this situation, any laser fluence and repeating rate, even, of course, those that weren't reflected in the experimental results that were acquired, might be predicted by the neural network as to the feature depth.

Metal additive manufacturing (AM) is a newly developed method for the economically advantageous production of low-volume complicated mechanisms. With AM, it is possible to create elevated superalloys, such as steel materials Inconel 718 and 316, which are difficult to make using conventional methods, into custom-shaped structures. AM constructions with fewer welds may last longer in the corrosive, elevated environment of nuclear reactors than conventionally produced ones [22]. The LPBF process is being used for AM of elevated metals, which have melting temperatures exceeding 130°C [23]. Keyhole and absence of fusion tiny pores may form in the AM metal due to the inherent characteristics of LPBF [24]. An AM construction must undergo nondestructive assessment (NDE) to find any potential defects prior to deployment in a nuclear reactor. In order to monitor the health of AM structures, in-service NDE inspections are required because it is unknown how the long-term behavior of AM metals in a nuclear environment will behave. In theory, high-resolution imaging of metals can be provided via X-ray computed tomography (XCT) [25]. However, XCT is restricted to penetration depths of around a centimeter and requires a symmetric body of revolution forms. Additionally, the relationship between structural size and XCT imaging resolution is inverse. Neutron tomography allows for deeper penetration, although the activation of the metal could be problematic. Although ultrasonic NDE can be applied to any shape or size of the structure, it needs to make direct contact with the structure's surface. Ultrasonic NDE poses issues because AM metals have an abrasive surface finish. Porosity and subsurface flaws in AM materials are less susceptible to conventional ultrasonic testing techniques than partial penetration infusing faults [26]. Pore detection is more successful with laser ultrasonic inspection [27]. However, surface roughness makes it difficult to detect defects accurately. Because inductive probes are non-contact and durable in a hostile environment, eddy current imaging is commonly employed in nuclear reactor in-service NDE applications. Although temperature and surface roughness may have an impact on readings, eddy current testing has the ability to find underlying pores [28]. Additionally, raster screening with a single probe is often required for Eddy's current imaging, which takes time. In comparison to single probe imaging, imaging with capacitive probe arrays has an inferior resolution.

5.2.1 DL-Based Predictive Visualization of Fiber Laser Machining

An extremely popular non-contact manufacturing method in both academia and industry is fiber laser materials processing. However, flaws like cracks and striations are typically seen on the cut material's surface, therefore modeling the interaction of light and matter is of particular importance. Since laser machining is a highly nonlinear process, it might be difficult to represent it using equation-based methods (such as finite element modeling), especially since many effects' physical causes are still unclear [29]. Recent research has shown that DL is capable of mimicking femtosecond laser machining [30]. A data-driven process is used in DL modeling, where the model is dynamically constructed from experimental data. As a result, DL offers a great chance to simulate as yet unrecognized laser machining effects, which helps with parameter optimization or even offers new information and knowledge. In this project, ten laser scan rates (15–24 m/min) were used to cut stainless steel sheets that were 2 mm thick and provided by TRUMPF Laser UK. Using a 5x objective, microscope images of the cut sides were captured, providing a total of about 50 mm of photos all along cut axis for each speed. There were three deep-learning modeling strategies investigated. First, the laser scan speed of the laser-machined surfaces was determined from microscope pictures using a CNN, which may be used to determine the category that corresponds to an image. The CNN was able to classify the ten categories with an accuracy rate of 98%, proving that there were distinct visual elements associated with each scan speed. Second, given a range of laser scan speeds, the appearance of the laser-cut sample was predicted using a cGAN, a neural network that can produce images.

A promising technique for post-manufacturing NDE of AM structures is pulsed thermography (PT). With such a flash, a material surface is temporarily heated, and a fast-frame infrared camera is used to capture time-resolved thermal images. Thermal image analysis is used to extract material flaws. The advantages of this technology include the speedy processing of large sample regions obtained in a single image and one-sided non-contact measurements. The ability to recognize minute details in thermal pictures is affected by edge blurring, uneven specimen heating, and image thermal noise patterns. Algorithms for image processing should be used to increase the effectiveness of

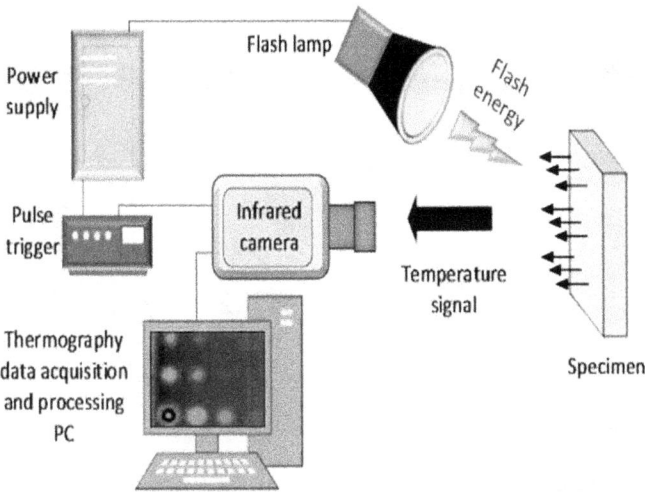

Figure 5.6 PIT imaging system.

flaw detection. Defect identification can be improved using conventional image processing techniques like sharpening, histogram equalization, and Gaussian filtering, but each one is only capable of addressing a single issue (Figure 5.6).

To enhance the quality of thermal imaging, methods like Pulsed Phase Thermography (PPT) [31] have recently been developed. The PPT uses a one-dimensional Discrete Fourier Transform (DFT) to extract phase information before transforming time-domain data into frequency spectra. Uneven heating can be lessened with the PPT approach; however, it is susceptible to high-frequency noise.

Thermographic Signal Reconstruction (TSR) is a technique that can significantly improve thermography frame resolution and make faults in the temporal and spatial domains more visible [32]. However, because so many frames need to be examined, the TSR method is ineffective. A high-energy capacitor discharging through a white light ash lamp is started by the pulse trigger in the PIT imaging technique, depositing a heat pulse on the sample surface. The megapixel quick frame IR camera synchronized to pulse trigger begins to record specimen blackbody radiation as heat permeates the entire material. Time-resolved pictures of the temperature field T on the specimen surface can be produced using the photo counts (x, y, t). An image analyzer unit is then used to perform additional processing on the stack of thermography images that have been recorded for material flaw detection, reconstruction, and compression methods. An image of the experimental PIT imaging lab setup can be seen in Figure 5.7. A thermal pulse of 6.4 kJ/2 ms is delivered to the specimen surface by a white light flash lamp that is supplied by a Balcar ASYM 6400 capacitor source. A lens is used to collimate flashlight. The material's surface is commonly covered with

Figure 5.7 Experimental PIT imaging system laboratory setup.

repositionable graphite black paint to enhance heat absorption. A high-speed >mid-wave IR (3–5 m) cooled detectors array camera (FLIR X8501sc) with NETD D 20mK is used for imaging. At full window size, the X8501sc model has a maximal spatial resolution of 1280×1024 pixels at a frame rate of 181 Hz.

5.2.2 Using ML, the Effects of Laser Energy Uncertainty on Temperature Changes in Directed Energy Deposition

A special capability for creating intricate 3D items from computer-aided design models is AM technology. Directed Energy Deposition (DED), one of the various metallic AM methods, is a versatile method suited to repair operations. In this process, metallic powder is accumulated and melted using a concentrated heat source. DED is increasingly being employed in sectors including aerospace [33] and bio-design [34]. There are several obstacles preventing the widespread adoption of AM in many industrial areas, including the high initial investment prices and inconsistent quality of printed parts. Different sources of uncertainty can have an impact on temperature evolution throughout the DED process, which is a crucial parameter to define the final microstructure [35]. Properties of the raw material, processing parameters, and environmental factors are sources of uncertainty [36]. In a streamlined scenario, each parameter change will necessitate a time-consuming process in order to anticipate the temperature field of the DED process [37]. Thousands of simulations will be necessary to discover the best parameters, which is extremely expensive in terms of computer and labor (Table 5.1).

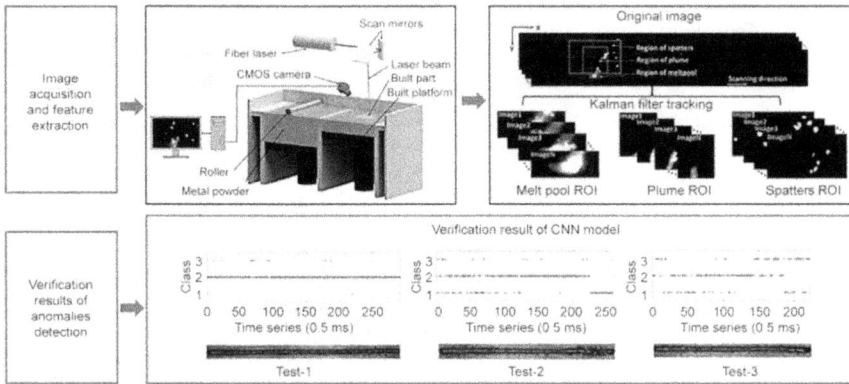

Figure 5.8 Pulsed infrared thermal imaging system.

5.2.3 ML-Based Challenges

Table 5.1 Challenges of ML in LBM

Challenge	Description
Lack of data labeling experience	The majority of NN application cases include supervised learning, which uses outputs as learning objectives. However, labeling data can be exceedingly challenging at times. How, for instance, may the various items in Figure 5.8 be correctly categorized as melt pool, plume, or spatters? In other words, the analyst's in-depth understanding of the welding process is extensively relied upon in these decisions. The advancement of NNs in the AM region will be severely hampered by such a reliance. In other words, extensive collaboration between specialists in computer science and material science is necessary for the widespread use of NNs to AM.
Underfitting and overfitting	An ML algorithm's main objective is to have a good generalization ability, which is a gauge of how well the algorithm can forecast results from previously unexplored data. However, the underfitting or overfitting issue is a reason why NN algorithms don't function well. The model is overfit, which occurs when the NN algorithm attempts to fit each individual piece of data from the training dataset. This makes the model especially susceptible to noise or outliers. On the other side, underfitting happens when the NN algorithm cannot find a reasonable relationship between the points in the training set. Dropout [38] and regularization [39] are two strategies to prevent overfitting and underfitting.

(Continued)

Table 5.1 (Continued) Challenges of ML in LBM

Challenge	Description
Lack of expertise in identifying desirable traits	The qualities of AM parts may be significantly impacted by a number of processing parameters while being only marginally impacted by others. On the other hand, with a small dataset, a large number of input features could quickly overfit the model. Making ensuring the NN algorithm is using a proper collection of characteristics is essential as a result. Researchers can greatly profit from feature engineering, a sort of preprocessing of incoming data. It has two distinct features: Finding the best features from the available inputs is the goal of feature selection. People may select factors such as "hatch distance,", "layer thickness" and "laser power," as having the most impact on a part's properties. In this case, the selection criteria are based on the researchers' prior knowledge and experience with AM, namely their ability to conduct in-depth investigations into the mechanics underlying the AM process rather than merely repeating studies. Utilizing statistical software to conduct quantitative analysis is another beneficial strategy. Some commonly used parameters in statistical science are listed below. When it is close to 1/1, the Pearson correlation coefficient, a useful tool for evaluating the relationship between the two features, shows a strong positive or negative relationship between the various inputs. Another metric to gauge the nonlinear link between two features is the Kendall rank correlation coefficient. To compare one numerical property to the others, mathematicians employ scatter matrices. It is feasible to learn which attributes have a stronger correlation with the desired property through the calculation of these parameters.

5.3 ROLE OF AI IN LBM

Robotics, natural language processing, ML, and image processing are just a few of the disciplines where AI is being actively researched. AI and ML have long been seen as "black-art" technologies, and there is frequently a lack of reliable proof to convince business that these techniques will consistently generate a return on investment. The effectiveness of ML algorithms is also significantly influenced by the preferences and experience of the developer. As a result, AI's performance in industrial applications has been inconsistent. Contrarily, the methodical discipline of industrial AI focuses on developing, analyzing, and using various ML algorithms for long-term performance in industrial applications [40]. It serves as a bridge between academic research findings in AI to industry practitioners and serves as a systematic technique and discipline to offer solutions for industrial applications. The ability of computers to carry out mental processes similar to those of humans, such as perception, reasoning, learning, and problem-solving, is known as AI [41]. The term "AI" first appeared in the 1950s with the development of the perceptron, a neural network (NN) structure designed in 1956 to mimic a human neural system by employing a weighted sum of inputs [42]. Despite its roots in human learning mechanisms and widespread belief that human-like cognitive AI was well within reach, the perceptron's progress was hampered by its inability to understand even the most basic logic and prohibitive computational complexity. This led to a significant decline in interest in supporting AI-related research, which signaled the beginning of the first AI winter. The development of expert systems, which required specialized handiwork and produced a vast list of "if-then" rules, was crucial in enabling the rebound from the first AI winter. Many individuals believed that using expert knowledge to construct AI was the most effective method due to this conventional wisdom.

The main driving forces and requirements in the bulk of everyday production industries across all industries are to meet quality, throughput, and cost targets while guaranteeing a safe working environment for everyone. Meeting these goals has become increasingly difficult as a result of the numerous requirements imposed by the complexity of the processes and products, the shifting expectations and preferences of the customer base, and the competitive pressure to maintain profitability. Positively, the challenging business environment that most manufacturers are currently experiencing offers an opportunity for the distinctive advantages that AI has over traditional tools and methods. Particularly, the routine task of problem-solving, which entails seeking out underlying causes, is well suited for the application of AI tools that are accomplished of seeing and categorizing multivariate, nonlinear patterns in performance and operational data that are obscure to the plant engineer. Nowadays, machines, controllers, environmental sensors, labor records, etc. produce enormous amounts of continuously generated data. These categories could be used to group the data:

1. Environmental information gathered from ambient sensors, such as humidity and temperature.
2. Process information gathered from sensors on process equipment or stations, such as coolant temperatures for machining and grinding, power, and heat treat thermal.
3. Measuring system or check data from product quality inspections, such as product diameter, form, and balance.
4. Effective production data is recorded in controller systems, such as relapsed time or timestamps of each part in each operation station.
5. Machine downtime, idle time, starvation/blockage, and shift scheduling.

This vast amount of data includes previously unheard-of opportunities for pattern discovery that can offer vital suggestions for solving complex problems and provide an additional description of the physiological importance of parameters to other physical components of a system or process. In addition to its ability to analyze high-dimensional data, AI enables the transformation of vast quantities of complex industrial data, which is becoming routine in today's industries, into meaningful and informative information.

Expert systems turned out to be too expensive to develop and maintain because of their inherent complexity, bringing in the second AI winter and shortening the lifespan of the knowledge-based AI age. Specifically, the advancement of DL and upgrades to the infrastructure for sensing and processing that made it possible to build ML/DL models. There are numerous techniques to represent the quality of weld products, depending on the various monitoring objectives. In order to achieve ideal weld seam penetration, for instance, the gathered signal data will be used to fine-tune the welding process variables or make real-time predictions. Extensive research has been conducted on online or offline or online monitoring of welding problems with the aim of eliminating weld defects. In order to ensure that the laser beam is always focused on the center of the welding gap, seam tracking technology is used. By utilizing feedback control technology for the welding process, we can effectively mitigate the influence of unknown interference on weld quality. Accurate signals can be obtained by relying on trustworthy monitoring equipment, which is the cornerstone of efficient monitoring. Modern data analysis techniques can greatly increase the monitoring capacity. Laser welding monitoring has steadily made use of AI, particularly ML. All areas of welding monitoring, including depth mining of welding signal data, predicting weld shape, and categorizing welding problems, have produced positive results. As a result, real-time laser welding monitoring has greatly advanced thanks to the use of cutting-edge monitoring tools and efficient AI-based techniques. The use of real-time surveillance in industrial manufacturing is currently constrained by a few issues. For example, the X-ray imaging system, one of the most cutting-edge monitoring instruments, is prohibitively expensive and has health risks that make it unsuitable for use in a plant. Due to the inefficiency of the multi-sensor data hybrid technique, data mining, feature extraction, and data fusion all take an inordinate amount of time. Use of DL is relatively new, and maximizing the benefits of data mining and self-renewing learning capabilities are the primary obstacles to lowering data processing time and improving monitoring accuracy. Insufficient background in data labeling.

AI has been used in a number of recent works due to its enormous potential for enhancing and automating AM operations. AI is being used in one application as a tool for process parameter analysis and labeling. Huang and Li [43] use AI algorithms for post-print analysis to investigate how manufacturing variables affect the reliability of a product's quality printout. For instance, they train ML algorithms like SVM and random forests on microscopic high-resolution images to determine that the interplay between part position and post-chamber pressure drop is an important factor in their process. As they dug deeper into the mystery of the one-of-a-kind case of downward-facing surface mistakes. Charles et al. [44] examined the extent following the construction. To acquire a deeper understanding of the effects of the process factors, they then apply regression and the ANOVA test. In their research, the automated method developed by Gobert et al. to identify breaks in CT images after they have been printed is used [45]. The discovered regions are then used to categorize images shot on-location with a 36-megapixel DSLR camera in layers. The team then creates an SVM that can identify aberrant layers with 85% precision. Lack of fusion porosity in powder bed fusion is attributed to insufficient overlap between melt pools, as stated by Tang et al. [46]. To calculate the extent of the overlap between lateral neighboring melt pools, they employ models in which the hatch spacing, layer thickness, and melt pool bridge area are input. The percentage of unmelted material is then forecast. Values for the energy density that are in agreement with the literature can be calculated by using the guideline values of relevant measurements using publicly available data. Afrasiabi et al. [47] exploited particle splitting to enhance the LPBF smoothed particle-hydrodynamics simulation's dynamic refining process. Incorporating spatial adaptability into their strategy, they were able to significantly reduce computational time. The test scenarios of recreating a liquid droplet and laser spot welding demonstrated only minor variations in expected breadth and depth when compared to state-of-the-art outcomes. To illustrate the effect of the stochastic powder bed on the operating window, Rausch et al. [48] employ a computer simulation. They demonstrate how the Lattice Boltzmann method-based mesoscopic methodology influences gas pores and single and multilayer bonding flaws in powder. Even while such settings show how AI can enable aid with deep analysis of enormous volumes of data, they are only accessible to a small group of users due to the need for a collection of specialized measurement gear (CT, Light Microscope, etc.).

Thus, modern studies rely on AI systems fed information from simpler sensor and camera arrays. The study [49] uses the temperature field images captured by an infrared thermal camera to spot errors in ongoing laser sintering processes. By using principal component analysis, they are able to extract and compress features that exist on a variety of dimensions, from the single to the multi (PCA). Using an SVM trained with seven of the PCA machineries, they achieve 88.9% accuracy in categorizing parameters as out-of-range. The study [50], using a melt pool analysis, locates areas of enhanced light emission in drift layers over 18 different build sections (BP). Using heat transfer features, they build a k-means model with an accuracy of 85%. When it comes to AI for AM, there is also a variant that use simpler camera setups to keep an eye on the printed layers by acquiring axial or coaxial views as opposed to, say, focusing on the melt pool. The study [51] build a web-based quality check that uses Bayesian categorization. By including a USB digital camera for picture capturing, their method enables in-situ diagnosis of issues like fusion flaws. By employing manually defined regional data, the built Bayesian model achieves an F1-Score of 86.03 when classifying images as either good or poor. To capture real-time images of their laser-cutting melting system, [52] use a high-speed camera. Using a CNN that also considers the temporal information of the retrieved characteristics, they are able to achieve a 99.7% accuracy for categorizing overheating, irregularity, normal, and balling. In their work on classification, [53] describe multiple forms of mistake. They used professionally tagged images to train an AlexNet and a multiscale convolutional neural network (MsCNN) via transfer learning. They performed quite well on the MsCNN, achieving class accuracies between 72.7% and 99.7%. The technique proposed in [54] combines error classification with previous segmentation of problematic locations based on histograms. With a CNN, we can reliably categorize images as either super-elevated, incomplete, striping, or normal with an average accuracy of 98 (Table 5.2).

5.3.1 AI Challenges

Table 5.2 Challenges of AI in LBM

Challenge	Description
Robustness and reliability	An AI-based model's resilience is defined as the consistency of the model's output following unusual changes inside the input data. A hostile attacker, background noise, or the failure of other AI-based system components may be the source of this shift [55]. For instance, during telesurgery, an unidentified crash in the machine vision component may allow an HLI-based agent to mistakenly identify a patient's kidney as a bean. One of the several models Even though both models have performance comparable, the robust model is deployed first. This area is in its early stages since conventional processes like duplication and multi-version programming may not function with intelligent systems. There are other studies, like [56], that discuss the distinction between a learning model's accuracy and resilience. Theoretical understanding of robustness and dependability seems to be developing, and significant developments are likely to take place in this area. Because faulty ML models may have exhibited inadequate robustness, this issue may be difficult for HLI-based agents to solve. therefore, in practice, an HLI with this flaw is prone to errors.

(Continued)

Table 5.2 (Continued) Challenges of AI in LBM

Challenge	Description
Security	Secure approaches, such those described in [57], are frequently created using AI-based systems, but from another angle, it is clear that any piece of software, particularly learning systems, could be compromised by malevolent users. The challenging issue is a crucial subject that has gained a lot of attention in the design of intelligent systems. Take ant-based path planning as an illustration, where the pathfinding procedure is manipulated by hacking the pheromone update function. Security issues with AI could lead to There are a number of new issues that this study is unable to address; for additional details, see [58]. The next paragraph discusses a security aspect, concentrating on protecting data-driven ML.
Explainable AI	Explainable AI is a rapidly developing field that has a wide range of applications in industries like healthcare, transportation, and the military [59,60]. In this area, a learning model may be made explainable using a number of tools and procedures. With this skill, people may be able to trust the models' judgments from a variety of angles, including those involving bias and fairness issues, to name a couple. This implies that explain ability could influence how other problems, such fairness and trustworthiness.
Responsibility	Over the past century, people have operated machines. We can all agree that there are moral and legal concerns regarding the culpability of human agents in this case. However, given completely autonomous machines, this condition will alter. New issues arise because autonomous devices based on genetic algorithms, neural networks, and learning automata can no longer be predicted by human activities This problem becomes more crucial when designing HLI, as is explained in the paragraph after this one.
Predictability	Whether or whether an AI-based agent can forecast its decision in every circumstance is a significant problem that may not be solved [61,62]. We can determine whether or not future smart bots can be trusted and intelligent agents can be controlled directly as a result of this challenge. Given the predictability challenge's nature, tackling this problem is not simple. We highlight some of it in what follows. It should be emphasized that unpredictable behavior can be noticed in a Due to the nature of these types, an agent with reinforcement learning methods is necessary. [63–72] summarizes a few problems related to heavy vehicle dynamics, biomedical and innovative sustainable technologies that have an impact on how predictable AI-based agents are. New design and analysis tool with mathematical formulation appears to be a major factor in AI and ML-based technologies for sustainable manufacturing and development.

5.4 CONCLUSION

Applications of ML and AI can significantly improve the laser manufacturing industry. Although there has been considerable development, it is still a long way off before ML and AI solutions for automating AM are integrated into other production procedures or become a commodity for consumers. Design improvement, process optimization, design correlation, microstructural design, and defect reduction are all areas where AI may advance laser manufacturing. The present methodologies created for other processes can greatly assist AM; nonetheless, the key challenge at this time is the availability and dependability of the data required to train the ML algorithms. There are several different types of recent experimental info collected from the academic researchers or AM manufacturers, some of which are not openly accessible. Therefore, the creation of ML algorithms for laser production must take into account the importance of trustworthy data collecting, storage, and sharing. The observation and many experiments that have been or are presently being undertaken differ significantly from one another. As a result, the manufacturing sector must develop a location for data storage. For the data to have meaning and for the ML algorithms to work properly, the data production precondition must also be provided with the data.

REFERENCES

[1] Gjelaj, F., Berisha, B., "Optimization of turning process and cutting force using multi objective genetic algorithm," *Universal J. Mechan. Engineer.* 7, 64 (2019).

[2] McCarthy, J., *The inversion of functions defined by Turing machines Automata Studies (AM-34).* Princeton University Press, Princeton, NJ, 177–182 (1956).

[3] Turing, A. M., "Computing machinery and intelligence Computational," *Mach. Intelligent Mind* 49, 433–460 (1950).

[4] Tian, Y., Shin, Y. C., "Multiscale finite element modelling of silicon nitride ceramics undergoing laser-assisted machining," *J. Manuf. Sci. Eng.* 129, 287 (2007).

[5] Singh, G., et al., "Finite element modelling of laser-assisted machining of AISI D2 tool steel," *Mater. Manufact. Process* 28(4), 443–448 (2013).

[6] Caron, M., et al., "Deep clustering for unsupervised learning of visual features," In Proceedings of the European Conference on Computer Vision (ECCV) (2018).

[7] Lee, H., et al., "Lasers in additive manufacturing: A review," *Int. J. Precis. Eng. Manuf.-Green Tech.* 4(3), 307–322 (2017).

[8] Sparkes, M., Steen, W. M., ""Light" industry: An overview of the impact of lasers on manufacturing," *Adv. Laser Mater. Process.* 1–22 (2018).

[9] Taglia ferri, F., et al., "Study of the influences of laser parameters on laser assisted machining processes," *Procedia CIRP* 8, 170–175 (2013).

[10] Yang, J., et al., "Experimental investigation and 3D finite element prediction of the heat affected zone during laser assisted machining of Ti6Al4V alloy," *J. Mater. Process. Technol.* 210(15), 2215–2222 (2010).

[11] Sutton, R. S., Barto, A. G., *Reinforcement learning: An introduction.* MIT Press, Cambridge (2018).

[12] Arul Kumaran, K., et al., "Deep reinforcement learning: A brief survey," *IEEE Signal Process. Mag.* 34(6), 26–38 (2017).

[13] Dey, A., "Machine learning algorithms: A review," *Int. J. Comput. Sci. Info. Technol.* 7(3), 1174–1179 (2016).

[14] Rumelhart, D. E., et al., *Backpropagation: The basic theory. Backpropagation: Theory, architectures and applications.* Lawrence Erlbaum, London, 1–34 (1995).

[15] He, K., Zhang, X., Ren, S., Sun, J., "Deep residual learning for image recognition," In Proc. IEEE Conf. Computer Vison Pattern Recognition (CVPR), 770–778 (June 2016).

[16] Zhang, W., Li, R., Deng, H., Wang, L., Lin, W., Ji, S., Shen, D., "Deep convolutional neural networks for multi-modality isointense infant brain image segmentation," *Neuro Image* 108, 214–224 (March 2015).

[17] Zhang, W., Li, R., Zeng, T., Sun, Q., Kumar, S., Ye, J., Ji, S., "Deep model-based transfer and multi-task learning for biological image analysis," *IEEE Trans. Big Data* 6(2), 322–333 (June 2020).

[18] Vaswani, A., Shazeer, N., Parmar, N., Uszkoreit, J., Jones, L., Gomez, A. N., Kaiser, L., Polosukhin, I., "Attention is all you need," (2017), arXiv:1706.03762. [Online]. Available: https://arxiv.org/abs/1706.03762.

[19] Devlin, M.-W. Chang, K. Lee, K. Toutanova, "BERT: Pre-training of deep bidirectional transformers for language understanding," (2018), arXiv:1810.04805 [Online]. Available: https://arxiv.org/abs/1810.04805

[20] Xiong, Y., Guo, L., Huang, Y., Chen, L., "Intelligent thermal control strategy based on reinforcement learning for space telescope," *J. Thermo Phys. Heat Transf.* 34(1), 37–44 (Jan 2020).

[21] Luo, W., Li, Y., Urtasun, R., Zemel, R., "Understanding the effective receptive field in deep convolutional neural networks," (2017), arXiv:1701.04128 [Online]. Available: https://arxiv.org/abs/1701.04128.

[22] Lou, X., Gandy, D., "Advanced manufacturing for nuclear energy," *JOM* 71, 2834–2836 (June 2019).

[23] Khairallah, S. A., Anderson, A. T., Rubenchik, A., King, W. E., "Laser powder-bed fusion additive manufacturing: Physics of complex melt flow and formation mechanisms of pores, spatter, and denudation zones," *Acta Mater.* 108, 36–45 (Apr. 2016).

[24] Hensley, C., Sisco, K., Beauchamp, S., Godfrey, A., Rezayat, H., McFalls, T., Galicki, D., List, F., Carver, K., Stover, C., Gandy, D. W., Babu, S. S., "Qualification pathways for additively manufactured components for nuclear applications," *J. Nucl. Mater.* 548 (May 2021), Art. no. 152846.

[25] Du Plessis, A., Le Roux, S. G., Booysen, G., Els, J., "Quality control of a laser additive manufactured medical implant by X-ray tomography," *3D Printing Additive Manuf.* 3(3), 175–182 (Sep 2016).

[26] Zhou, Z. G., Sun, G. K., "New progress of the study and application of advanced ultrasonic testing technology," *J. Mech. Eng.* 53(22), 1–10 (Dec 2017).

[27] Millon, C., Vanhoye, A., Obaton, A. F., Penot, J. D., "Development of laser ultrasonics inspection for online monitoring of additive manufacturing," *Weld World* 62(3), 65–661 (Feb 2018).

[28] Du, W., Bai, Q., Wang, Y., Zhang, B., "Eddy current detection of subsurface defects for additive/subtractive hybrid manufacturing," *Int.J. Adv. Manuf. Technol.* 95(9–12), 3185–3195 (Apr 2018).

[29] Ermolaev, G. V., et al., "Mathematical modelling of striation formation in oxygen laser cutting of mild steel," *J. Phys. D: Appl. Phys.* 39(19), 4236 (2006).

[30] Heath, D. J., et al., "Machine learning for 3D simulated visualization of laser machining," *Optics Express* 26(17), 21574–21584 (2018).

[31] D'Accardi, E., Palano, F., Tamborrino, R., Palumbo, D., Tatì, A., Terzi, R., Galietti, U., "Pulsed phase thermography approach for the characterization of delaminations in CFRP and comparison to phased array ultrasonic testing," *Journal of Nondestructive Evaluation* (1) (2019).

[32] Shepard, S. M., Beemer, M. F., "Advances in thermographic signal reconstruction," In Proc. SPIE 9485, Thermosense: Thermal Infrared Applications XXXVII (2015), Vol. 9485.

[33] Kamal, M., Rizza, G., "Design for metal additive manufacturing for aerospace applications," *Additive Manufact. Aerospace Indus.* 67–86 (2019).

[34] Sarraf, M., Rezvani Ghomi, E., Alipour, S., Ramakrishna, S., Liana Sukiman, N., "A state-of-the-art review of the fabrication and characteristics of titanium and its alloys for biomedical applications," *Bio-Design Manufact.* 1–25 (2021).

[35] Jardin, R. T., Tchuindjang, J. T., Duchêne, L., Tran, H. S., Hashemi, N., Carrus, R., Mertens, A., Habraken, A. M., "Thermal histories and microstructures in direct energy deposition of a high speed steel thick deposit," *Mater. Lett.* 236, 42–45 (2019).

[36] Jardin, R. T., Tuninetti, V., Tchuindjang, J. T., Hashemi, N., Carrus, R., Mertens, A., Duchêne, L., Tran, H. S., Habraken, A. M., "Sensitivity analysis in the modeling of a high-speed, steel, thin wall produced by directed energy deposition," *Metals* 10(11), 1554 (2020).

[37] Zheng, B., Haley, J. C., Yang, N., Yee, J., Terrassa, K. W., Zhou, Y., Schoenung, J. M., "On the evolution of microstructure and defect control in 316L SS components fabricated via directed energy deposition," *Mater. Sci. Engineer. A*, 764, 138243 (2019).

[38] Ng, A. Y., "Feature selection, L1 vs. L2 regularization, and rotational invariance," In Proceedings of the 21st International Conference on Machine Learning July 4–8, 2004, Banff, AB, Canada (2004).

[39] Srivastava, N., et al., "Dropout: A simple way to prevent neural networks from overfitting," *J. Machine Learn. Res.* 15(1), 1929–1958 (2014).

[40] Wang, L., "From intelligence science to intelligent manufacturing," *Engineering* 5(4), 615–618 (2019).

[41] Chui, L., Kamalnath, V., McCarthy, B., "An executive's guide to AI, McKinsey" (2018), https://www.mckinsey. com/business-functions/mckinsey-analytics/ our-insights/an-executives-guide-to-ai.

[42] Cardon, D., Cointet, J. P., Mazières, A., "Neurons spike back: The invention of inductive machines and the artificial intelligence controversy," *Reseaux* 5(211), 173–220 (2018).

[43] Huang, D., Li, H., "A machine learning guided investigation of quality repeatability in metal laser powder bed fusion additive manufacturing," *Mater. Des.* 203, 109606 (2021).

[44] Charles, A., Elkaseer, A., Thijs, L., Scholz, S., "Dimensional errors due to overhanging features in laser powder bed fusion parts made of Ti-6Al-4V," *Appl. Sci.* 10, 2416 (2020).

[45] Gobert, C., Reutzel, E., Petrich, J., Nassar, A., Phoha, S., "Application of supervised machine learning for defect detection during metallic powder bed fusion additive manufacturing using high resolution imaging," *Additive Manufact.* 21, 517–528 (2018).

[46] Tang, M., Pistorius, P. C., Beuth, J. L., "Prediction of lack-of-fusion porosity for powder bed fusion," *Additive Manufact.* 14, 39–48. Elsevier BV. (2017).

[47] Afrasiabi, M., Lüthi, C., Bambach, M., Wegener, K., "Multi-resolution SPH simulation of a laser powder bed fusion additive manufacturing process," *Appl. Sci.* 11(7), 2962. MDPI AG. (2021).

[48] Rausch, A., Küng, V., Pobel, C., Markl, M., Körner, C., "Predictive Simulation of process windows for powder bed fusion additive manufacturing: Influence of the powder bulk density," *Materials* 10(10), 1117. MDPI AG. (2017).

[49] Chen, Z., Zong, X., Shi, J., Zhang, X., "Online monitoring based on temperature field features and prediction model for selective laser sintering process," *Appl. Sci.* 8, 2383 (2018).

[50] Yadav, P., Singh, V., Joffre, T., Rigo, O., Arvieu, C., Guen, E., Lacoste, E., "Inline drift detection using monitoring systems and machine learning in selective laser melting," *Advanced Engineering Materials* 22, 2000660 (2020).

[51] Aminzadeh, M., Kurfess, T., "Online quality inspection using Bayesian classification in powder-bed additive manufacturing from high-resolution visual camera images," *J. Intelligent Manufact.* 30, 2505–2523 (2018).

[52] Zhang, Y., Soon, H., Ye, D., Fuh, J., Zhu, K., "Powder-bed fusion process monitoring by machine vision with hybrid convolutional neural networks," *IEEE Trans. Indust. Informat.* 16, 5769–5779 (2020).

[53] Scime, L., Beuth, J., "A multi-scale convolutional neural network for autonomous anomaly detection and classification in a laser powder bed fusion additive manufacturing process," *Additive Manufact.* 24, 273–286 (2018).

[54] Shi, B., Chen, Z., "A layer-wise multi-defect detection system for powder bed monitoring: Lighting strategy for imaging, adaptive segmentation and classification," *Materials & Design* 210, 110035 (2021).

[55] Hanif, M. A., Khalid, F., Putra, R. V. W., Rehman, S., Shafique, M., "Robust machine learning systems: Reliability and security for deep neural networks," In Proceedings of the 2018 IEEE 24th International Symposium on On-Line Testing and Robust System Design (IOLTS), Platja d'Aro, Spain, 257–260 (2–4 July 2018).

[56] Qayyum, A., Qadir, J., Bilal, M., Al-Fuqaha, A., "Secure and robust machine learning for healthcare: A survey," *IEEE Rev. Biomed. Engineer.* 14, 156–180 (2020).

[57] Pawar, S. N., Bichkar, R. S., "Genetic algorithm with variable length chromosomes for network intrusion detection," *Int. J. Automation Computer* 12, 337–342 (2015).

[58] Yampolskiy, R. V., *Artificial intelligence safety and security.* Boca Raton, FL, USA, CRC Press (2018).

[59] Adadi, A., Berrada, M., *Explainable AI for healthcare: From black box to interpretable models. In Embedded Systems and Artificial Intelligence.* Springer, Berlin/Heidelberg, Germany, 327–337 (2020).

[60] Gade, K., Geyik, S. C., Kenthapadi, K., Mithal, V., Taly, A., "Explainable AI in industry," In Proceedings of the 25th ACM SIGKDD International Conference on Knowledge Discovery & Data Mining, Anchorage, AK, USA, 3203–3204 (4–8 August 2019).

[61] Yampolskiy, R. V., Unpredictability of AI. arXiv 2019, arXiv:1905.13053.

[62] Musiolik, G., "Predictability of AI decisions. In Analyzing future applications of AI, sensors, and robotics in society; IGI Global: Hershey, PA, USA," 17–28 (2021).

[63] Bhoi, S., Prasad, A., Kumar, A., Sarkar, R. B., Mahto, B., Meena, C. S., Pandey, C., "Experimental study to evaluate the wear performance of UHMWPE and XLPE material for orthopedics application," *Bioengineering* 9, 676 (2022). 10.3390/bioengineering9110676

[64] Patil, P. P., Sharma, S. C., Jaiswal, H., Kumar, A., "Modeling influence of tube material on vibration based EMMFS using ANFIS," *Procedia Mater. Sci.* 6, 1097–1103 (2014).

[65] Kumar, A., Rana, S., Gori, Y., Sharma, N. K., "Thermal contact conductance prediction using FEM based computational techniques," In *Advanced computational methods in mechanical and materials engineering.* CRC Press, Boca Raton, FL, USA, 183–220 (2021). ISBN 9781032052915.

[66] Kumar, A., Datta, A., Kumar, A., "Recent advancements and future trends in next-generation materials for biomedical applications," In *Advanced materials for biomedical applications.* Kumar, A., Gori, Y., Kumar, A., Meena, C. S., Dutt, N., Eds. CRC Press, Boca Raton, FL, USA, Chapter 1; 1–19 (2022).

[67] Prasad, A., Chakraborty, G., Kumar, A., "Bio-based environmentally benign polymeric resorbable materials for orthopedic fixation applications," In *Advanced materials for biomedical applications.* Kumar, A., Gori, Y., Kumar, A., Meena, C. S., Dutt, N., Eds. CRC Press, Boca Raton, FL, USA, Chapter 15; 251–266 (2022).

[68] Datta, A., Kumar, A., Kumar, A., Kumar, A., Singh, V. P., "Advanced materials in biological implants and surgical tools," In *Advanced Materials for Biomedical Applications.* Kumar, A., Gori, Y., Kumar, A., Meena, C. S., Dutt, N., Eds. CRC Press, Boca Raton, FL, USA, Chapter 2; 21–43 (2022).

[69] Kumar, A., Gangwar, A. K. S., Kumar, A., Meena, C. S., Singh, V. P., Dutt, N., Prasad, A., Gori, Y., "Biomedical Study of Femur Bone Fracture and healing," In *Advanced materials for biomedical applications.* Kumar, A., Gori, Y., Kumar, A., Meena, C. S., Dutt, N., Eds. CRC Press, Boca Raton, FL, USA, Chapter 14; 235–250 (2022).

[70] Gangwar, A. K. S., Rao, P. S., Kumar, A., "Bio-mechanical design and analysis of femur bone," *Materials Today: Proceedings* 44(Part 1), 2179–2187 (2021), ISSN 2214-7853. 10.1016/j.matpr.2020.12.282

[71] Gangwar, A. K. S., Rao, P. S., Kumar, A., Patil, P. P., "Design and analysis of femur bone: BioMechanical aspects," *J. Critical Rev.* 6(4), 133–139 (2019), ISSN-2394-5125.

[72] Meena, C. S., Kumar, A., Jain, S., Rehman, A. U., Mishra, S., Sharma, N. K., Bajaj, M., Shafiq, M., Eldin, E. T., "Innovation in green building sector for sustainable future," *Energies* 15, 6631 (2022). 10.3390/en15186631

Chapter 6

Application of Laser Technology in the Mechanical and Machine Manufacturing Industry

Aayush Pathak, Abhishek Kumar, Avinash Kumar, Ashwani Kumar, and Francis Luther King M

CONTENTS

6.1 Introduction 108
    6.1.1 Application of Lasers 108
        6.1.1.1 In Defence Sector 109
        6.1.1.2 In Telecommunication 109
        6.1.1.3 In the Medical Industry 109
        6.1.1.4 In the Manufacturing Industry 110
    6.1.2 Advantages of Lasers..110
        6.1.2.1 Extreme Precision 110
        6.1.2.2 Automated..110
        6.1.2.3 Energy Efficient 111
        6.1.2.4 No Tool Wear 111
    6.1.3 Limitations of Lasers..111
6.2 Current Tools and Techniques Used in Mechanical and Machine
    Industry...111
    6.2.1 Laser Cutting, Drilling, and Piercing.............................112
        6.2.1.1 Laser Cutting...112
        6.2.1.2 Laser Drilling and Piercing.............................113
        6.2.1.3 Limitations of the Existing Technology.............115
        6.2.1.4 Other Applications of Laser Cutting 117
    6.2.2 Laser Welding 118
        6.2.2.1 Typical Setup of Laser Welding........................120
        6.2.2.2 Applications of Laser Welding in General.........120
        6.2.2.3 Applications of Laser Welding in the Machine
                Industry 122
        6.2.2.4 Limitations to the Existing Technology.............122
    6.2.3 Additive Manufacturing 122
        6.2.3.1 Laser's Role in Additive Manufacturing............122
    6.2.4 Various Techniques Used for Additive Manufacturing 123
        6.2.4.1 Laser Powder Bed Fusion 123
        6.2.4.2 Laser Engineered Net Shaping........................126
        6.2.4.3 VAT Photopolymerization................................128

DOI: 10.1201/9781003402398-6

6.2.4.4   Sheet Lamination/Laminated Object
Manufacturing.. 129
6.2.4.5   Two-Photon Polymerisation 130
6.2.4.6   Layered Manufacturing Issues in General 130
6.2.4.7   Surface Treatment by Laser.............................. 131
6.2.4.8   Laser Cleaning.. 134
6.2.4.9   Medical Devices Manufactured Using Laser...... 137
6.2.5   Suitable Alternatives to Improve the Existing
Technology... 140
6.2.5.1   Laser Cutting.. 140
6.2.5.2   Laser Welding 142
6.2.5.3   Laser Drilling 143
6.2.5.4   Additive Manufacturing 144
6.2.5.5   For Surface Treatments 145
6.3   Summary, Conclusion, and Future Challenges........................... 146
References.. 150

6.1 INTRODUCTION

The word LASER stands for "light amplification by stimulated emission of radiation". In Layman's terms, one can regard lasers as nothing more than a torch or a flashlight; energy goes in the form of electricity and we get the output as light. But the concept involved in laser generation is way different from that of light coming out of a torch or a flashlight.

For now, we can say that there are three major differences when we compare lasers to our ordinary light and these are as follows:

1. Ordinary light is less constrained than laser light.
2. The flashlight beam is white in colour which is a combination of different colours of light mixed together whereas the laser light always exhibits a single colour.
3. In contrast to laser light waves which are well aligned to each other, the light waves in ordinary light are arranged in a random manner.

The narrow beam of a laser light is a boon for many industrial as well as day-to-day applications that we see around us almost every day. The CD or DVD writer in your PCs uses the laser technology to read the information present on them. You must have gone to a grocery store and must have seen the barcodes of the products being read using lasers. It can even cut through metals and has many advantages over the regular machining processes.

6.1.1 Application of Lasers

Almost every field uses laser technology in some capacity whether it is for telecommunication purposes or even in the defence sector.

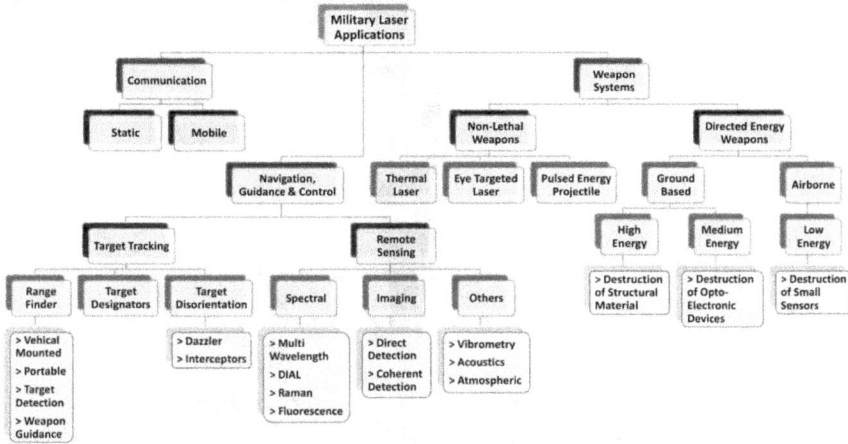

Figure 6.1 Military laser applications [3].

6.1.1.1 In Defence Sector

Today, a wide variety of lasers with various wavelengths, operating conditions, and power capacities are available. By acting in a variety of capacities, including sensing devices, weather controllers, indicators, rangefinders, data relays, target designators, and directed energy weapons (see Figure 6.1) [1]. The laser systems used for direct energy weapons can prove to be lethal and can cause damage on a very large scale. The most accurate and precise way to eliminate moving targets over a thousand miles is to use a high-powered laser beam [2].

6.1.1.2 In Telecommunication

The major use of lasers in communication purposes is by optical fibres. Every time someone uses telephone for communication or even while connecting to the internet, data is transmitted by optical fibres. These optical fibre cables have a lot of advantages as compared other telecommunications mediums, faster internet speed, and very less signal attenuation to name a few.

6.1.1.3 In the Medical Industry

Lasers beams having the order of wavelength of 1 μm can invade the eye and is successfully used to treat Cataracts. Other applications include kidney stone removal without any surgery. Thread-like glass strands are sent through the urethra to the kidneys and then intense laser pulses are sent through these strands which eventually break up the kidney stone into small pieces making it easier for its passage through the urethra.

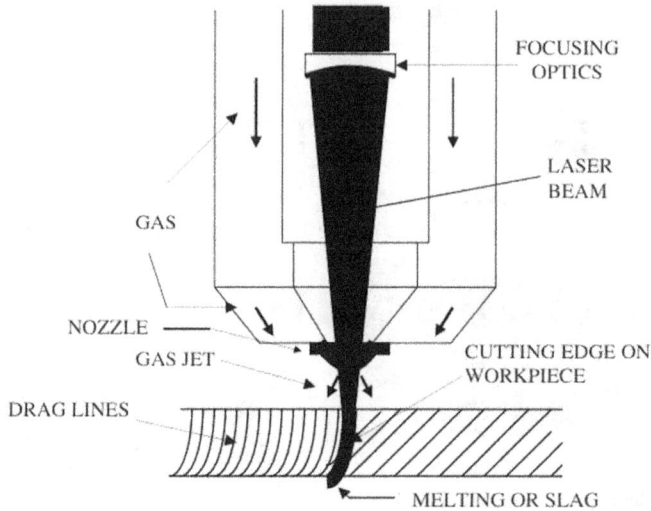

Figure 6.2 Laser cutting schematic [5].

6.1.1.4 In the Manufacturing Industry

Through melting or the vaporisation processes, laser beams offer a concentrated heat source that can be used to "drill" holes, connect materials, shape, and cut metals [4]. The two major types for lasers used for the cutting process are the carbon dioxide laser and the Nd:YAG laser (see Figure 6.2).

6.1.2 Advantages of Lasers

6.1.2.1 Extreme Precision

Laser machining is extremely precise and can even reach a dimensional accuracy of 0.0005 inches. This is due to the following reasons:

- As laser beams emit coherent photons and this coherently intense beam of photons can easily penetrate through almost any material.
- High intricacy ensures less damage to the material leading to a detailed finish and desired design.
- Laser beams can be tuned very easily according to the type of material needed to perform the operation on, this also provides flexibility on the energy usage.

6.1.2.2 Automated

The laser systems can be connected to a computer numeric control (CNC), and with the use of computer programmes, we can programme the laser system setup to perform the exact same operations as prescribed in the computer programme. No matter how large the work, CNC programming easily changes the cutting's speed and intensity to produce the same precise, accurate outcome.

6.1.2.3 Energy Efficient

Although lasers require a high amount of energy to operate, but the time required to perform an operation on a material is very less when compared to the conventional operations which eventually leads to less energy usage than other techniques in a long run. Nowadays, almost every laser beam system uses robots as they are reliable which allows cost-effective manufacturing.

6.1.2.4 No Tool Wear

As compared to any conventional machining process, there is no tool wear in laser systems as there is no direct contact between the workpiece and the laser; it is just the heat that melts and vaporises the material.

6.1.3 Limitations of Lasers

- The laser systems are very expensive as compared to other machines required to perform the same operation.
- In the medical field, laser surgery involves risks including scarring, infection, and skin and eye related issues.
- Machines equipped with laser systems require highly skilled workers to pay a regular attention while a particular operation is being carried out because a slight mistake in adjusting the intensity or the temperature can lead to serious damage.
- As the parts involved in such laser systems are very intricate in nature, a regular and frequent maintenance of such systems is required.

6.2 CURRENT TOOLS AND TECHNIQUES USED IN MECHANICAL AND MACHINE INDUSTRY

Laser is nowadays one of the most important and useful tool when we talk about the manufacturing and machine industry. Primarily used for cutting sheet metals lasers can now be used to drill, weld, and engrave, just to name a few. As the technology is advancing day-by-day and the research is flourishing at a very high pace, lasers have become more efficient in terms of both energy and time consumption.

6.2.1 Laser Cutting, Drilling, and Piercing

In your childhood, you all have used a magnifying glass on a bright sunny day to burn a piece of paper. That is exactly the concept of laser cutting and drilling; the only difference is that the source is laser here instead of ordinary light.

The advantages of using laser in cutting, drilling, or piercing operations are as follows:

- As compared to Hot Jet processing or any other thermal cutting process where a round corner is obtained, we can obtain a sharp corner using laser.
- The cut requires no post-processing and requires no further treatment or cleaning.
- It is faster than other conventional processes.
- As there is no direct contact between the workpiece and the laser, no tool wear occurs.
- There is no directional dependence with cutting orientation.
- There is no or very less noise level.
- The process is highly flexible and easy in nature as there are mainly soft tool changes, i.e. the only changes are in the programming that controls the tool trajectory.
- Laser operations are compatible with almost every material. They can be brittle, hard, soft, non-conducting, or conducting; only a few reflective materials like copper and aluminium can create some problems but these can also be cut very easily by proper beam control.

6.2.1.1 Laser Cutting

The setup used for laser cutting is shown in Figure 6.3.

The major components of such a setup includes:

- Laser itself along with a shutter control.
- Focusing optics.
- Beam guidance train.
- Means to move the workpiece and the beam relative to each other.

The beam is then diverted into a beam dump, which also serves as a calorimeter, by the shutter, which is often a retractable mirror. A solenoid or pneumatic piston quickly removes the mirror when the beam is needed. The beam then travels to the beam guidance train, which guides it to centre on an optical focusing device. There can be two types of focusing optics either the reflective optics or the transmissive optics.

The transmissive ones are generally made of GaAs, ZnSe, or CdTe whereas the reflective ones are made on-axis spherical mirrors or parabolic off-axis mirrors. The coaxial gas jet then emerges from a nozzle after the

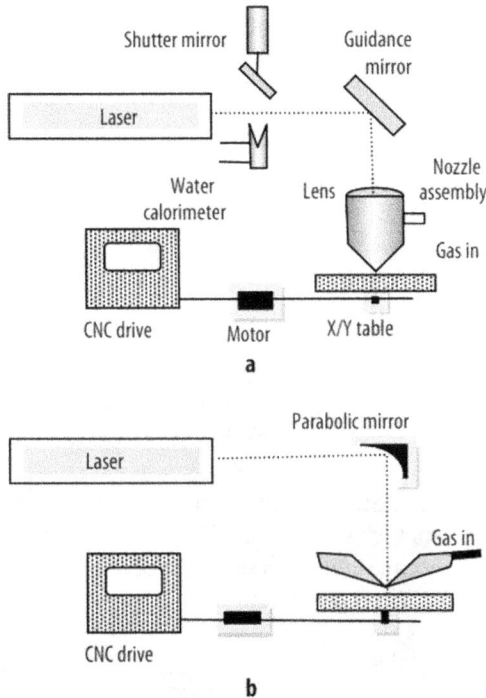

Figure 6.3 General arrangement for laser cutting. **a**: Using transmissive optics, and **b**: Using reflective optics. CNC—computer numerical control [6].

concentrated beam has passed through it. An "Air Knife" is also used and it blows diagonally over the optic train's departure resulting in the removal of spatter and smoke. While cutting, the laser vaporises a hole in the material, which is then passed through to create the cut. Because there is no open-sided hole, drilling and piercing are distinct from cutting in a line.

6.2.1.2 Laser Drilling and Piercing

A ruby laser was employed in some of the early high-powered laser studies, and its power was determined by how many Gillette razor blades it could pierce in a single shot. The ability to pierce steel sheets with a single pulse of light was considered exceptional in those days. Today, there are several uses for precisely laser-drilled holes. Boundary layer film cooling is one of the primary uses in jet engine parts like combustion chambers and turbine blades.

Laser drilling is advantageous due to the following reasons:

- Decrease or management of the taper.
- Removal of spatter.

- Decrease or removal of the layer that has resolidified on the hole wall.
- Repeatability.
- Drilling a hole through a coating.

Due to the necessity of laser drilling as an industrial operation, there are numerous ways to produce quick, excellent holes with strong repeatability. They include:

1. **Single pulse drilling:** The hole is made and finished in one pulse (see Figure 6.4).
2. **Percussion drilling:** Without moving the workpiece or the beam, single or many shots are performed (see Figure 6.5).
3. **Trepanning:** To form a cut, the beam is rotated around the circumference of the hole.
4. **Helical trepanning:** Starting close to the centre of the hole, revolving around its perimeter, spirally drilling into the workpiece with each rotation, and occasionally changing the focal point to follow the hole's base downward (see Figure 6.6).

Using these methods, holes can be made:

- Normal to the workpiece: The conventional method.
- At an angle to the workpiece surface: Very useful for turbine blades.
- As blind holes.

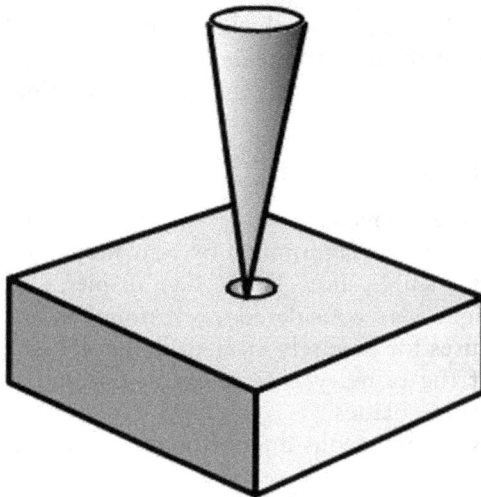

Figure 6.4 Single pulse drilling schematic [6].

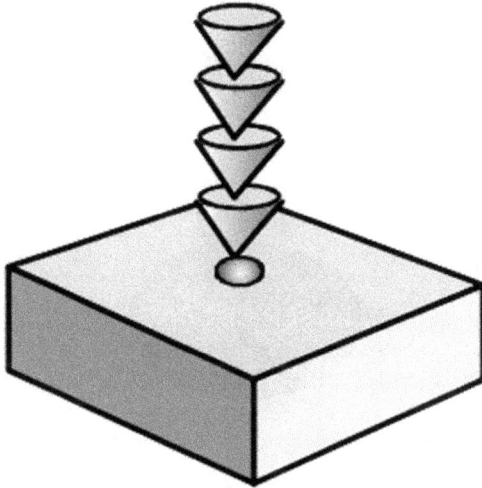

Figure 6.5 Percussion drilling schematic [6].

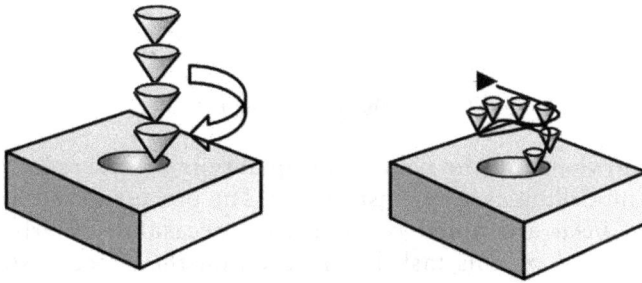

Figure 6.6 Trepanning schematic and helical trepanning schematic [6].

6.2.1.3 Limitations of the Existing Technology

There are a few concerns and issues affecting the laser cutting according to [7] and are mentioned in Figure 6.7.

The first strategy is connected with advancements in laser beam monitoring and control. Power and spatial-density distributions can be regulated using a variety of real-time monitoring methods. Definitely advancements in quality and stability of the laser beam are required to improve the material processing quality. Even though laser beam monitoring can be improved, it does not take into account changes in workpiece material, plasma generation, or other interactions zone phenomena.

The second approach is regarding economics, i.e. altering beam conditions so that the energy consumption is optimum. Modelling can be categorised as the third such strategy for enhancing cut quality. The

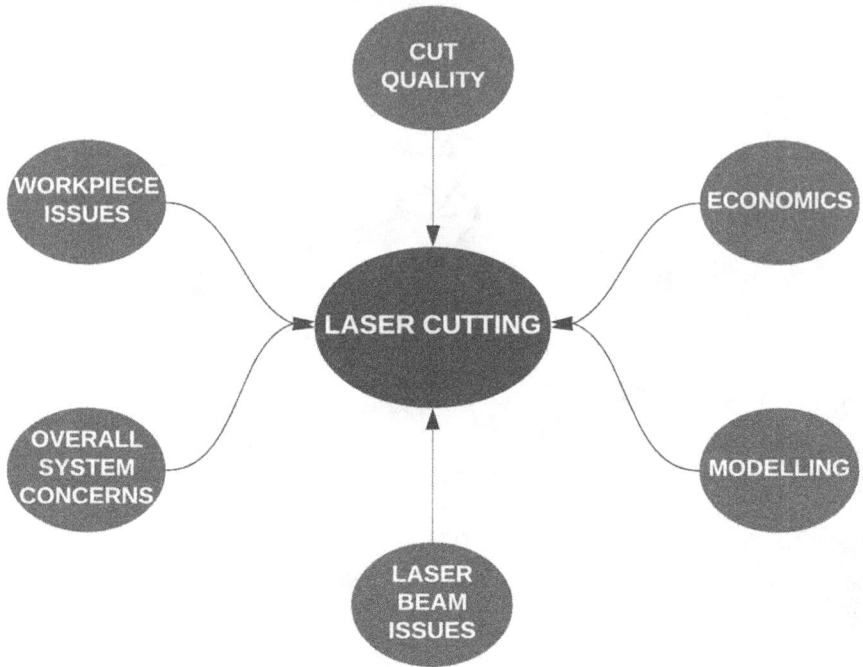

Figure 6.7 Concerns and issues affecting laser cutting [7].

quest of knowledge in the field of cut quality is considered as requiring the use of modelling as a key instrument. The process of modelling laser cutting has been attempted on numerous occasions. Nearly everyone who has worked on this task has focused on the process' steady-state behaviour.

The fourth such area of interest focuses on enhancing the workstation's or system's overall behaviour. This include changing the process parameters at certain points of profile change and contour programming. Today, they are often of a very good calibre and do not present as many challenges as some of the other problems.

Other limitations include:

- Although process parameters are calibrated and changed to produce the required cut quality, finding the ideal cutting circumstances may still prove to be a challenge.
- This process must be repeated if a different kind of material needs to be cut.
- According to [8] "Since the capital and operating costs for laser systems are large, high material removal rates, high dimensional accuracy, good surface quality, and a high degree of repeatability must be achieved to make these processes economically viable".

- Burnout is a risk when making sharp turns, thin slices, or anywhere else where the workpiece temperature is getting close to the ignition temperature. This can be avoided to a certain extent by
  - Using pulse power.
  - Increasing the power to match the speed.
  - Employing a water spray to surround the oxygen jet in a circle.
  - Spatter will result while piercing the sheet anywhere other than the edge.

The optics and hole quality are at risk because of this spatter.

6.2.1.4 Other Applications of Laser Cutting

The major application of lasers in an industry is for cutting purposes. Due to its speed and the superior cut quality it produces, the laser is cost effective since it decreases or completely eliminates the need for after-treatment, which results in significant cost savings. One of the cleanest and quickest profile cutting methods available today is laser cutting.

The simplicity of cutting profiles has made a number of intriguing design innovations possible. Cutting tabs and slots, for instance, can make assembly simpler. By making the tabs or slots various sizes, it might offer built-in mistake proofing and enable self-jigging for inexpensive fixtures. Another suggestion is to cut slots along the bends that are formed close to previously drilled holes to prevent the hole from being distorted during bending. Cutting a sequence of perforation slots along a bend line is an alternative to that, allowing for more accurate bending with less expensive tooling or even by hand.

- **Profile cutting:** This operation is basically used for gear manufacturing in the machine industry, gun parts, valve plates, medical components, gaskets, and many more. The dimensional accuracy acheived here is of a few micrometres with a fine finish.
- **Aerospace materials:** Lasers can cut brittle, hard ceramics like SiN ten times quicker than diamond saws can. Alloys of titanium cut in an inert atmosphere are used for airframe manufacturing. Aluminium alloys which require a laser power of greater than 1 KW, when compared to routing or blanking, cost savings of 60–70% have been observed.
- **Prototype car production:** Currently, robotically directed lasers are used to cut sunroofs for automobiles as an optional assembly step. On the assembled car, holes are also laser cut for left-hand-drive or right-hand-drive vehicles. General Motors used four Coherent General 300 W carbon dioxide lasers at its Delco Remy plant in Anderson, Indiana, in the United States, as the first automotive application in 1972 to cut ignition coils.

- **Cutting radioactive materials:** As the generator is out of the hot zone and only the workpiece, fixtures, and a few tiny optics are susceptible to contamination, working with radioactive materials with optical energy is significantly easier than working with other sources of energy. Additionally, optical energy can be sent over vast distances, so the task may not need to be restricted to particular locations. The absence of fume is one benefit of using a laser for cutting. There is some fume, but it is minimal in comparison to the other options. Thus, it is a compelling idea for dismantling and upkeep of nuclear power plants.

6.2.2 Laser Welding

One of the sources with the highest power density currently available to industry is the focused laser beam. The power density is comparable to that of an electron beam. These two procedures make up a portion of the cutting-edge high-energy processing technology. The table shown in Figure 6.8 compares the power density of various welding techniques. If the energy can be absorbed at these high-power densities, all materials will vaporise. Consequently, a hole is typically created during this type of welding through evaporation.

Less energy is used for unneeded heating or creating a HAZ or distortion, the greater the joining efficiency score. The energy is mostly generated at the high-resistance interface that needs to be welded, making resistance welding the ideal option in this regard. But it is clear that the electron beam and laser are once more in a league of their own. What are their performance characteristics in comparison to other processes, and can they be identified from one another? What kind of market expectations is it possible to

Process	Heat source intensity (W m^{-2})	Fusion zone profile	
Flux-shielded arc welding	5×10^6–10^8		
Gas-shielded arc welding	5×10^6–10^8		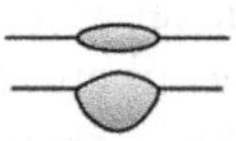 low / high
Plasma	5×10^6–10^{10}		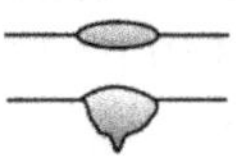 low / high
Laser or electron beam	10^{10}–10^{12}		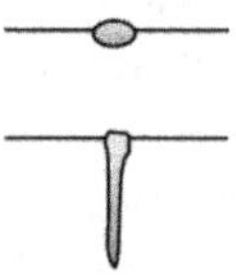 defocus / focus

Figure 6.8 Comparison of various welding techniques wrt their power densities [9].

Characteristic	Comment
High energy density – "keyhole" type weld	Less distortion
High processing speed	Cost-effective (if fully employed)
Rapid start/stop	Unlike arc processes
Welds at atmospheric pressure	Unlike electron beam welding
No X-rays generated	Unlike electron beam
No filler required (autogenous weld)	No flux cleaning
Narrow weld	Less distortion
Relatively little HAZ	Can weld near heat-sensitive materials
Very accurate welding possible	Can weld thin to thick materials
Good weld bead profile	No clean-up necessary
No beam wander in magnetic field	Unlike electron beam
Little or no contamination	Depends only on gas shrouding
Relatively little evaporation loss of volatile components	Advantages with Mg and Li alloys
Difficult materials can sometimes be welded	General advantage
Relatively easy to automate	General feature of laser processing
Laser can be time-shared	General feature of laser processing

Figure 6.9 Main characteristics of laser welding [9].

predict for laser welding? Is it a ploy or a present? The table shown in Figure 6.9 represents the main characteristics of laser welding.

The different welding techniques like TIG, resistance, ultrasonic, and laser welding can be compared with respect to the factors like rate of welding, operating costs, and equipment reliability, as shown in Figure 6.10. These tables demonstrate the laser's unique potential as a high-quality, high-speed welding tool. The laser is especially advantageous when welding heat-sensitive parts, such as narrow diaphragms on bigger frames, pistons constructed with washers in place, and heart pacemakers.

Quality	Laser	Electron beam	TIG	Resistance	Ultrasonic
Rate	✓	✓	×	✓	×
Low heat input	✓	✓	×	✓	✓
Narrow HAZ	✓	✓	×		✓
Weld bead appearance	✓	✓	×		✓
Simple fixturing	✓	×	×		
Equipment reliability	✓		✓	✓	
Deep penetration	×	✓		×	
Welding in air	✓	×		✓	
Weld magnetic materials	✓	×	✓	✓	✓
Weld reflective material	×	✓	✓	✓	✓
Weld heat-sensitive material	✓	✓	×	×	✓
Joint access	✓			×	×
Environment, noise, fume	✓	✓	×	×	×
Equipment costs	×	×	✓		
Operating costs	–	–	–	–	–

✓ point of merit, × point of disadvantage

Figure 6.10 Comparison of welding processes [9].

6.2.2.1 Typical Setup of Laser Welding

Welding uses a precisely focused beam to achieve the penetration, just like laser cutting. The only exception will be when the seam is to be welded had a fluctuating gap or was difficult to track, in which case it would be simpler and more reliable to utilise a broader beam. In this instance, the competition from a plasma process should be taken into account once the beam has been defocused. The idea behind laser beam welding is that it only works when an atom's electrons are stimulated by being exposed to energy. After some time has passed, it eventually returns to its ground energy level and emits a light photon. We get a high-energy focused laser beam as a result of the excited emission of radiation increasing the concentration of this released photon. The welding equipment is initially set up in the desired area (between the two metal parts to be joined). Later, a high-voltage power supply is used to operate on the laser apparatus. The laser is directed by the lens into the desired welding area. During the welding process, computer-aided manufacturing (CAM) is utilised to regulate the speed of the workpiece table and laser. The machine's flash lamp is turned on, and light photons are released. The atoms of ruby crystals absorb the energy of light photons, which excites electrons to higher energy levels. A photon of light is released when they reach its ground state or low-energy level. Once more stimulating the atom's electrons, this light photon creates two more photons. As the procedure goes on, a laser beam that is tightly concentrated is directed to the required spot to join numerous components.

The major components of a laser beam welding process are as follows:

- **Laser machine.**
- **CAM:** It is a type of CAM in which the welding procedure is carried out using a laser equipment and a computer. CAM executes all control operations during the welding process. This greatly accelerates the welding process.
- **CAD:** Computer-aided design is referred to as CAD. It is a piece of software that allows us to create welding jobs. Here, the workpiece is designed using a computer, along with how welding will be done on it.
- **Shielding gas:** During the welding process, shielding gas may be utilised to stop the workpiece from oxidising.
- **Power source:** The laser device generates a laser beam using a high-voltage power source.

6.2.2.2 Applications of Laser Welding in General

Laser welding provides numerous advantages. Some are listed as follows:

- A narrow HAZ, i.e. low distortion, and the potential for welding close to heat-sensitive materials, such as electronics and plastics, are the results.

- Aesthetically pleasing narrow fusion zone.
- High speed.
- Ease of monitoring and operating.

Numerous applications have resulted from these benefits, and as laser prices continue to drop and production engineers' understanding increases, the number of applications is expected to grow significantly in the future. Some applications are listed as follows:

- Bimetallic saw blade welding.
- Stamped muffler welding.
- Complicated form welding before pressing.
- Inside-out repair of nuclear boiler tubes.
- Strip welding used in continuous mills.
- One method being utilised on some heat exchangers and aviation parts is to weld a flat pack of two or more layers in an acceptable pattern before blowing the shape up with pressurised gas or hydraulically. Sheet metal items like heat exchangers and washing machines use this approach.
- There are many uses for cars such as welding of airbag sensors, gear wheels, torque converters, etc. [10].
- Polymer and plastic welding: Applications for plastic welding are numerous and growing. They include diving suits, tents, injection moulding, footwear, welding glasses, and wetsuits [11].
- For smart materials: Large-scale manufacturing of advanced materials is being carried out whose manufacturing can be automatic thus saving labour expenses [12].
- Structural panels (see Figure 6.11) [13].

Figure 6.11 Structural panel [13].

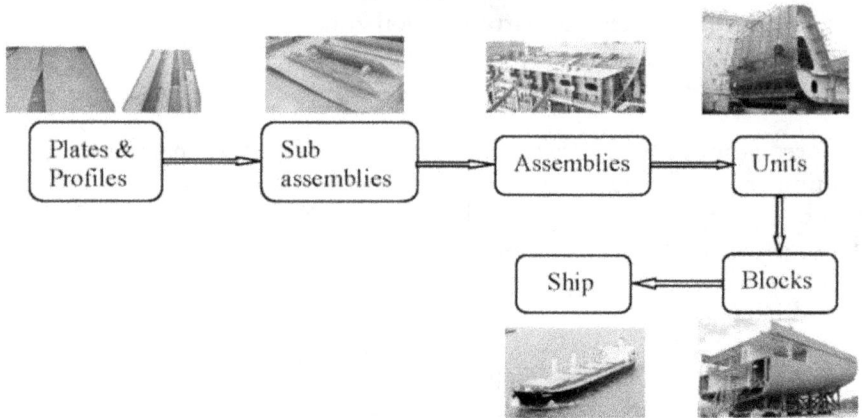

Figure 6.12 Shipbuilding stages [14].

6.2.2.3 *Applications of Laser Welding in the Machine Industry*

In the machine industry, laser welding can be used for:

- **Tool construction**
  With the advantage of controlling process parameters and high weld structure quality, it is used to make a variety of tools.
- **Shipbuilding**
  With laser beam welding, we can easily manufacture shipbuilding components like rudders, drive screws, and control because they require a particular amount of tolerance (see Figure 6.12). Extremely poor energy conversion efficiency exists in LBW.

6.2.2.4 *Limitations to the Existing Technology*

- The expensive equipment required for LBW is a significant drawback.
- The kind of application laser welding is used for is a limiting element.
- Some metals can fracture as a result of the quick cooling.
- To accomplish LBW, highly skilled workers are necessary.
- Extremely poor energy conversion efficiency exists in LBW.

6.2.3 Additive Manufacturing

6.2.3.1 *Laser's Role in Additive Manufacturing*

Lasers were discovered around 56 years ago [15] and inspired scientists, engineers, media, and even public in general. It was regarded as one of the 20th century's most important discoveries. Since then, the laser has advanced greatly. Laser plays an instrumental role for the manufacturing of a variety of products and helps to create new opportunities for the market. The applications of lasers have increased progressively in the recent times, ranging from cutting, drilling, and welding to, and in the last 5–10 years, in the field of additive manufacturing.

Additive manufacturing, generally abbreviated as AM and also called as 3D printing, is a bit different from our conventional manufacturing processes. The conventional manufacturing processes adapt a subtractive approach whereas AM uses a layer-by-layer deposition approach for manufacturing a particular product. AM has a lot of benefits as compared to the conventional approaches and have proved to be very effective in the defence and aerospace field [16].

6.2.4 Various Techniques Used for Additive Manufacturing

6.2.4.1 Laser Powder Bed Fusion

An energy source (a laser) is utilised to fuse the powder at certain points on a construction plate where the model specifies the desired geometry. A new layer of powder is placed when the first layer is finished, and the procedure is continued until a 3D part is created [17]. Laser powder bed fusion (L-PBF) has some alternate names also, as shown in Figure 6.13.

They consist of an energy delivery and a powder delivery system. The powder delivery system consists of a piston in order to supply the powder. In order to create the powder layer, a coater is also provided and a piston to hold the constructed component. The energy delivery system is made up of a laser (Ytterbium fibre laser having a wavelength of 1075 nm) and the scanning system that ensures the delivery of the laser spot at all positions on the building platform.

A continuous gas flow system (Ar or Nitrogen) in order to:

- Shield the component from oxygen.
- To remove metal fumes and spatter created while using laser.

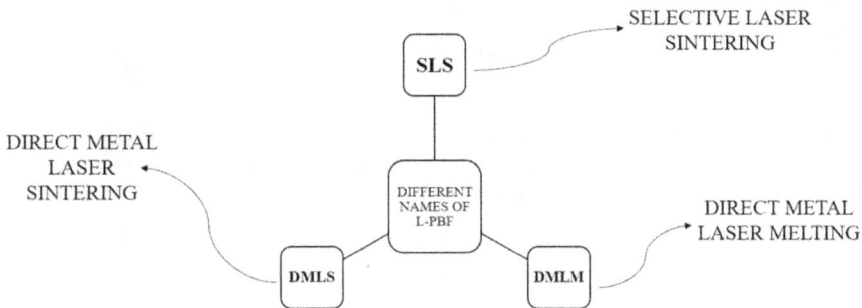

Figure 6.13 Other names of L-PBF [17].

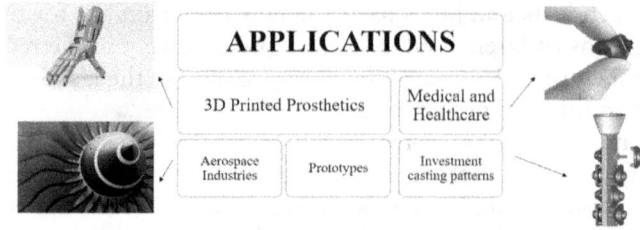

Figure 6.14 Various applications of L-PBF.

In some systems, a high-speed camera and a temperature controller are also used along with the laser system. The accuracy achieved is around ±0.05 to 0.25 mm. The powder materials that are generally used are different forms of **Polyamides** (like PA 12).

6.2.4.1.1 Advantages of L-PBF

- The processing speed is generally fast.
- Typically don't need a support structure.
- The parts produced exhibit high stiffness and strength.
- Complex geometries can be produced with ease.

6.2.4.1.2 Applications of L-PBF

L-PBF technique finds its application in various fields. Some of them are represented in Figure 6.14.

6.2.4.1.3 Limitations of the Existing Technology

- Equipment required for L-PBF is very costly.
- Power consumption is very high in this process.
- Distortion and fissures can result from shrinkage.
- The surface of the parts appears to be grainy without post-processing.

6.2.4.1.4 Defects in the L-PBF Process

The defects that generally occur in the L-PBF process are classified into four major categories viz. geometry and dimension, surface finish, micro-structure, and mechanical properties [18].

6.2.4.1.5 Geometry and Dimensional Defects

There are two parameters that contribute to such defects.

- **Staircase effect:** The "staircase effect" is a 3D printing phenomenon where the layer lines on the surface of the pieces become clearly apparent and give the impression of a staircase [19]. Thus, the name.
- **Machine error conditions/parameters:** There are two machine parameters that lead to dimensional and geometrical inaccuracy.
  - A defective laser focus on the build platform also called as laser positioning error.
  - Incorrect vertical movement of the manufacturing platform also called as platform movement defect.

6.2.4.1.6 Defects Due to Weak Mechanical Properties

The following are the reasons for weak mechanical properties of the additively manufactured products:

- **Fractures:** Pertaining to the spread of cracks, which causes the material or component to fail [20]. There are two types of fracture:
  - **Inter-granular fracture.**
  - **Trans-granular fracture.**
- **Cracks:** It is basically regarded as a void in a material which acts as stress concentration sites.
- **Holes:** They are very similar to cracks and may be present at the inclusion sites.
- **Porosity:** Porosity is nothing but the measure of the amount of spaces present in a material and is basically represented as a fraction of the volume of the voids to the total volume and is expressed as a percentage [21].

6.2.4.1.7 Microstructural Defects

Microstructural defects arise due to the following factors:

- **Anisotropy:** Layer orientation and scan direction affect the isotropic property of the additively manufactured part. Tensile strength and elongation are impacted by the scanning direction [18]. To print isotropic parts, a multi-directional scanning mode is required.
- **Heterogeneity.**
- **Microstructural porosity/poor density:** Poor density occurs due to the following reasons:
  1. **Laser mode:** When compared to continuous wave, pulsed mode consolidates metallic powders at less average power which in turn increases the density of the consolidated component.
  2. **Balling:** Due to surface tension and a lack of wetting ability with the prior layer, melted material solidifies into spheres, which is directly related to the properties of the melt pool [22].

3. **Powder size:** If the powder size is large, it will require high-power density for melting, resulting in high porosity. That is why it is recommended to use powder whose size is less than 100 μm.
4. **Warping:** Possibly between two layers or at the line separating the support layer from the component layer [21].
5. **Unfused powder:** Unfused powder basically occurs due to insufficient melting or due to the overlap between two melted tracks which are adjacent to each other.

6.2.4.1.8 Defects Due to Surface Quality and Integrity

The defects due to surface quality include (see Figure 6.15) [23]:

- **Powder deposition:** Powder deposition should be uniform in order to achieve good surface integrity.
- **Substrate quality:** The quality of surface of the deposited layer's succeeding layers is significantly influenced by the layer's roughness. This is due to the fact that faults such as increased porosity and decreased bonding between the two adjacent layers resulting from a rougher surface.
- **Staircase effect.**
- **Surface pits:** Due to rapid solidification, sometimes particles spherical in orientation are created and are broken down by the recoating blades resulting in pit formation on the build surface.

6.2.4.2 Laser Engineered Net Shaping

It is basically used for the additive manufacturing of metallic materials. This method is very similar to selective laser melting except for some parameters [24]. In laser engineered net shaping (LENS), the material and energy source are introduced simultaneously while it is operating.

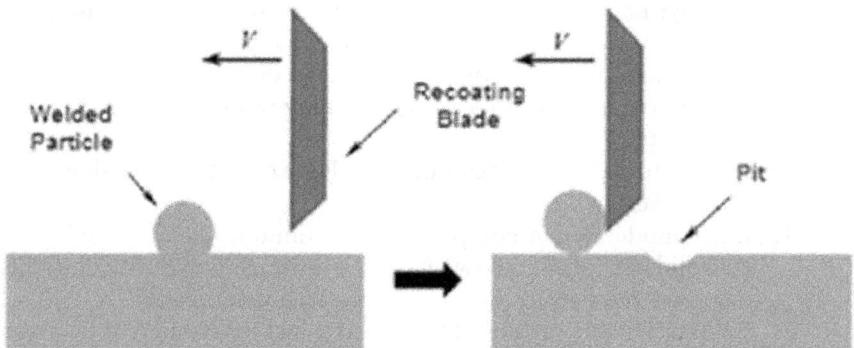

Figure 6.15 Defects caused by recoating blades [23].

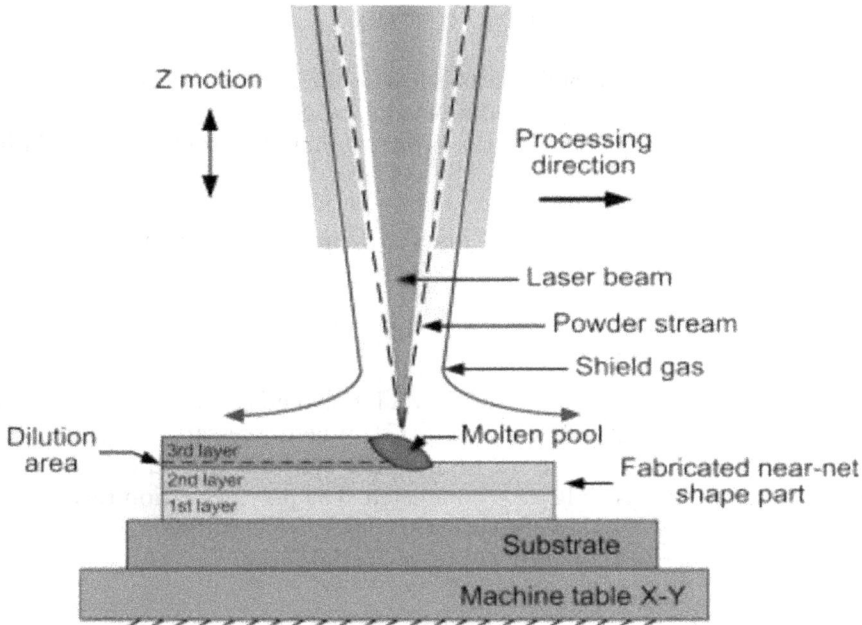

Figure 6.16 LENS process [25].

The material can be utilised either as a filament or as a powder. Gas or air pressure is used to push the feedstock upto the laser tip and onto the build surface. Then with the help of the focused laser tip, this filament or powder layer is melted by adjusting the laser energy density. To avoid the part distortion during the process, it requires overhangs and support structures (see Figure 6.16) [25].

With LENS, we can achieve a dimensional accuracy of 50–100 μm. The materials which are processed with this technique are generally metals and metal alloys like nickel alloys, ceramics, titanium alloys, steel, etc.

6.2.4.2.1 Applications of LENS

- Medical and healthcare.
- Repairing machine parts.

6.2.4.2.2 Advantages of LENS

- Mechanical strength of the build parts is high.
- Complex and intricate geometries can be easily made by LENS.
- High productivity.
- Utilised to precisely restore worn-out parts.

6.2.4.2.3 Limitations of the Existing Technology

- The density of the build parts is low so they cannot be used at certain areas where the density requirements are high.
- The surface roughness of the products manufactured using LENS is quite high.
- High capital cost.
- The build resolution is very low for the parts/products manufactured by LENS.

6.2.4.3 VAT Photopolymerization

A liquid photopolymer resin used in VAT polymerisation is used to build the model layer by layer (see Figure 6.17). When necessary, an ultraviolet (UV) laser is utilised to cure the resin, and a platform is used to lower the object as each successive layer is completed. The polymerisation reaction is triggered and manifested by the curing UV laser, which subsequently creates chains of cross-linked polymer to create a solid resin.

6.2.4.3.1 Steps Involved in the Process

- **Step-1:** Before the deposition of a new layer, the build platform is lowered down by a certain amount (equal to the layer thickness).
- **Step-2:** The resin is cured layer by layer using a UV lamp. As the platform descends farther, new layers are added on top of the older ones.
- **Step-3:** To create a smooth resin foundation in order to produce the following layer, some machines employ a blade that travels between layers.
- **Step-4:** After the deposition is completed, the VAT is drained off the resin and the manufactured product is removed.

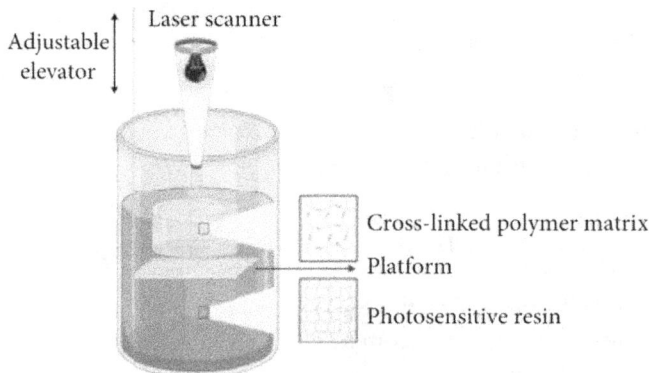

Figure 6.17 VAT photopolymerization [26].

6.2.4.3.2 Limitations of the Existing Technology

- The products made by VAT photopolymerization have less strength and durability.
- If they are not post-cured, resins might distort with time.
- This technique is relatively costly.
- The post-processing time for such products is very high.

6.2.4.4 Sheet Lamination/Laminated Object Manufacturing

The basic idea behind this method is to gradually adhere small layers of material covered with glue. The supply roll is used to get the sheet material. Once the laminated sheet is in position, the part's outline is cut out of the sheet using a laser. The excess material outside of the part boundary is then cross-hatched by the beam, transforming it in the form of a support structure having a rectangular boundary that stays in place. The build platform is then moved up and a hot roller moves over it, attaching the new layer to the one that came before it in a single circular motion. The laser beam is then used to cut the part's outline. This keeps happening until the complete section is created. After one layer is deposited, the platform is lowered by a height equal to the thickness of one layer to be deposited and the process is repeated until we get the desired thickness of the product [27].

6.2.4.4.1 Applications of Laminated Object Manufacturing

- **Design:** For proof of concept, visualisation, verification of CAD models, and marketing.
- **Engineering analysis and planning:** Stress and flow analysis, planning surgical operations, and for prosthetics and implants.
- **Manufacturing:** For casting, moulding, etc., to name a few.

6.2.4.4.2 Limitations to the Existing Technology

- Products manufactured by laminated object manufacturing (LOM) have a poor surface finish.
- Interior structural problems and undercuts.
- Bonding between the adjacent layers is not up to the mark for some materials.
- Unlike other additive manufacturing processes, this process is not good for products that have intricate and complex geometries.
- Unused material that results from interior hollow areas requires special techniques and skilled workers to be removed.
- Finishing cost is generally high for LOM as compared to other AM processes.

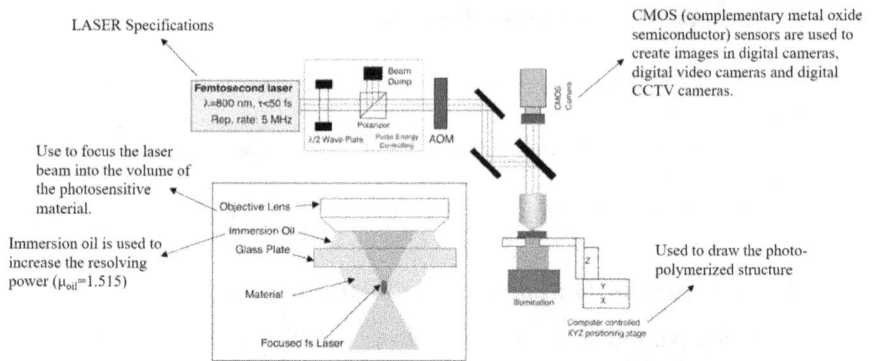

Figure 6.18 Two-photon polymerisation schematic [29].

6.2.4.5 Two-Photon Polymerisation

6.2.4.5.1 Basic Physics Involved

The concurrent absorption of two photons material which is photosensitive in nature results in the non-linear optical process known as two-photon polymerisation [28,29]. The so-called photo-initiators are activated during this process, which modifies the photosensitive material and causes a polymerisation. These develop into radicals that locally polymerise the resist. The dimensional accuracy is of the order of 100 nm. Two-photon polymerisation schematic is shown in Figure 6.18.

6.2.4.5.2 Limitations to the Existing Technology

- If we compare this technique to VAT photopolymerization, the absorption is less.
- Photobleaching is more prominent here.
- Dielectric breakdown can result in photodamage.
- Expensive.

6.2.4.6 Layered Manufacturing Issues in General

- General
  Although ideas for immediately printing genuine holographic images are being studied, building a 3D model layer by layer is currently the only commercially available rapid-build method. For instance, two-photon polymerisation, which uses femtosecond pulses that are not absorbed by the monomer but rather by two photons in a non-linear reaction, enables resolutions finer than the diffraction limit as well as the casting of holographic images as objects in three dimensions as opposed to layering.

- **Layer thickness issue**
  We can reduce the staircasing effect by reducing the thickness of the layers being deposited but for this, we have to compromise with the computational time. The optimum approach is to have a changeable step height dependent on the object's slope and the manufacturing process's capacity.
- **Accuracy**
  Although different additive manufacturing processes have different defects observed but distortion, curling, and shrinkage are common in all. These issues are dependent on the cooling rates and phase changes that the material undergoes.
- **Orientation of the parts**
  The layers built during a process are a function of the time required to perform the operation, i.e. the build time. Larger steps need not be particularly harmful because of the orientation. One can place the "stairs" on any face by changing the orientation. Therefore, the crucial faces ought to be constructed along the slicing plane.
- **Support structures**
  The overhanging parts need to be supported in various AM processes, especially the VAT photopolymerization process. Whereas some processes like the SLS process do not require any support structures because the support is provided by the powder bed itself. Support structures can also be used to prevent part deformation, especially in the case of large, thin fins (see Figure 6.19).

6.2.4.7 Surface Treatment by Laser

The laser has several distinct surface heating capabilities. The EM radiation emitted by the laser is absorbed by the first few atomic layers for materials which are opaque in nature (like metals). The plus point with laser is that it does not release jets of hot gas and causes no spillage of radiation outside

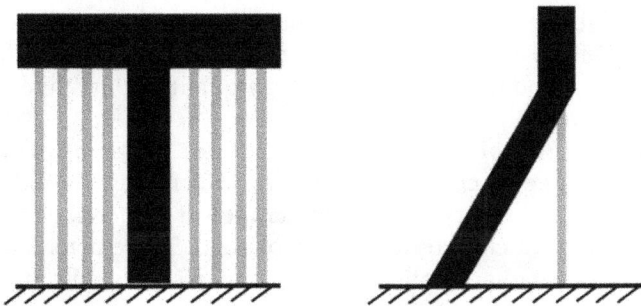

Figure 6.19 Support structure for better printability [30].

the focused beam trajectory. In actuality, the applied energy can be directed precisely to the required areas of the surface. The advantages of using laser for surface treatment purposes as compared to other alternative processes are mentioned as follows:

- Chemical sterility.
- Comparatively simple to automate.
- No or very less post-processing is required.
- The thermal penetration is quite controlled in nature.
- The thermal profile is also controlled.
- Non-contact process.

Because it appears to give the potential to preserve strategic materials or to enable superior components with ideal surface and bulk properties, surface treatment is a topic of great attention at the moment. These goals are true and attainable, but as we become older, we'll realise that only some regions of surfaces are susceptible to corrosion and wear and that larger areas don't need to be covered.

The surface treatment using laser finds its applications in areas like:

- **Surface melting:** For surface sealing, homogenisation, and refinement of material's microstructure.
- **Surface alloying:** For enhancing corrosion, wear, and aesthetic qualities.
- **Texturing of the surface:** The surface of the steel sheet rolls in temper mills is made dull which enhances the paint movement on the final surface. There are a lot of techniques to perform it (see Figure 6.20) [9,31,32]. The most common technique is to shot-blast the surface. However, this results in a random roughness that causes waviness in the painted surface. A consistent pattern of roughness can prevent this. With the help of the EDM or laser machining, we can achieve the uniform surface roughness levels. The laser used for such operation is either the carbon dioxide laser or the Nd:YAG laser. The patterned roughness must be extremely regular and uniform. Due to fluctuations inside the YAG rod, the Nd:YAG laser does not produce consistently high-quality spots like the EDM method does.
- **Scabbling:** Scabbling is basically used for the removal of surface made of stone or concrete. Before resurfacing or decommissioning, large, massive rotary claws can scrape off the concrete surface of waterworks, nuclear power plant, or road. These machines have the potential to damage concrete tanks unnecessarily and produce a lot of dust, which would be an issue in the nuclear example. Although fairly impressive if the method avoids the issues mentioned previously [33,34].

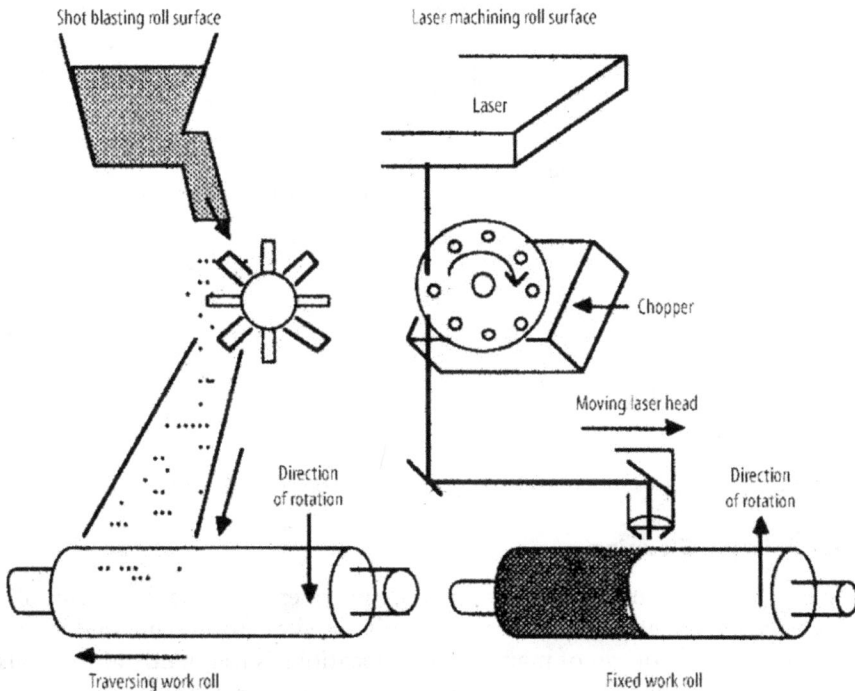

Figure 6.20 Texturing the surface using the shot-blast technique [9].

- **Cladding of the surface:** Two methods—preplaced powder deposition [35–39] and codeposition method [40–44]—can be used to supply the material that is to be deposited on a substrate. The clad material supply for these processes varies. In the first approach, a slurry is created by first combining the powder to be covered with certain types of adhesives (polyvinyl alcohol). This slurry is spread evenly over the substrate and left to dry and solidify (see Figure 6.21). This is done to prepare it to withstand the force of shielding gas and the laser's particle nature.

6.2.4.7.1 Limitations of the Existing Technology

- High installation and setup charges for the device.
- Cracking may occur if the build rate is too high.
- Constraints on large equipment size which means that the system cannot be moved.
- The cladding may experience stress cracking due to improper setup and power.
- Laser technology is still new, laser diode's availability and capabilities are currently relatively constrained.

(a)

(b)

Shielding gas

Powder or
Wire supply

Laser beam

Shielding
gas

Laser beam

Preplaced
material

Clad Layer

Melt pool

Melt pool

Scanning direction

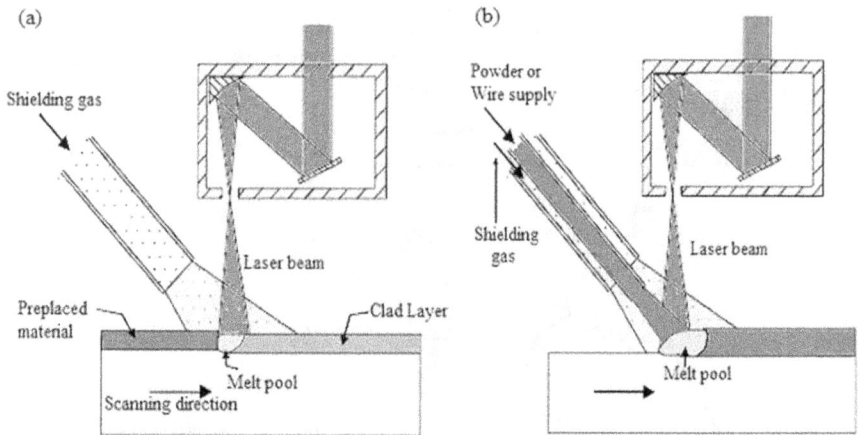

Figure 6.21 (a) Preplaced technique and (b) codeposition technique [45].

6.2.4.8 Laser Cleaning

The use of laser cleaning is becoming more prevalent, especially in fields like art conservation and the removal of minute dirt from semiconductors. While the specific origin of many laser applications is frequently ambiguous and the use of light to have a mechanical action on surfaces was recognised long before the advent of the laser, it is possible to unambiguously attribute laser cleaning to the following three originators.

The first originator was Arthur Schawlow (who was later awarded the Noble Prize in physics in 1981 for his study on laser spectroscopy) did his collaborative research with Charles Townes which led to the development of MASER (optical laser). The perception for building was that the laser was regarded as a "Solution searching for a problem" because potential military applications were being kept under wraps and were widely believed to be unfeasible. Schawlow was working to dispel this bias.

The second major event that led to the use of laser for cleaning purposes was in 1972. Following significant floods, the Italian Petroleum Institute requested John F. Asmus of the UCSD to explore laser holography in Venice in order to capture the city's deteriorating treasures. During this project, Asmus was asked to investigate the results of the interaction between a focussed ruby laser—which up until this point had been meant for holographic recording—and a stone statue covered in jewels. His Italian collaborators included conservator Giancarlo Calcagno. The darker hard coatings on the surface were discovered to have been removed from the surface in a selective manner, leaving the white stone beneath undamaged. Asmus eventually made his way back to the United States, where he started exploring laser cleaning of artwork. His research laid the groundwork for several potent approaches, including the use of pulsed ruby and pulsed Nd:YAG lasers [46–51].

Techniques were developed for removing tiny sub-micrometre waste particles from silicon wafers and other microelectronic devices by a team working with Susan Allen of FSU. This contained a proprietary method for applying a thin layer of a water–alcohol mixture that, due to the liquid's quick evaporation, considerably increased the laser cleaning efficiency [52,53].

6.2.4.8.1 Types of Laser Cleaning Techniques

- **Selective vaporisation:** According to Asmus, there are two main cleansing mechanisms. Cleaning took place in typical pulse mode at a comparatively low laser power due to the selective vaporisation of the surface impurities in comparison to the underlying material, which was largely unaffected. This in turn happened when the darker encrustation's absorption coefficient was sufficiently high to cause a temperature rise that was favourable for vaporisation, but the underlying material's absorption coefficient was too low to affect temperature rises to levels that permitted the absence of vaporisation which in turn prevented cracking.
- **Spallation:** Asmus coined a term "Ablation" which was later known as Spallation when the laser operated in a Q-switched mode (meaning that the duration of the pulse was around 10–20 ns) which was responsible for cleaning purposes (see Figure 6.22). At high laser flux values (nearly around 10^8 W/cm^2), even the surfaces that are highly reflective in nature absorb sufficient amount of energy which makes the surfaces reach their vaporisation temperature. This is due to the fact that at such high temperatures, the surface becomes ionised leading to high absorbtivity of laser intensities. As the target is protected from the laser with the help of a partially ionised ("plasma") vapour, the initial surface vaporisation ends. High pressures

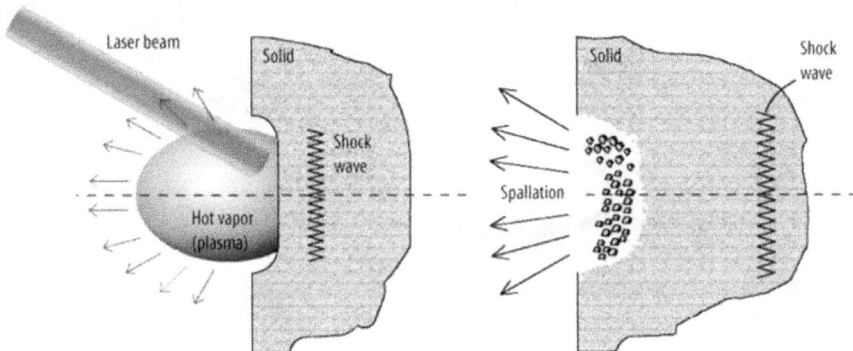

Figure 6.22 Spallation [9].

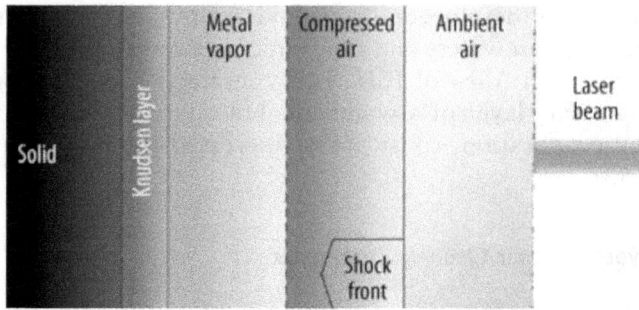

Figure 6.23 Evaporation pressure mechanism [54,55].

(of the order of 1–100 Kbar) are produced which results in a shock wave. This shock wave microscopically compresses the target material's surface. After the laser pulse, the material surface relaxes, the plasma spreads away from the surface, and a thin surface layer (1–100 μm) is eliminated, leading to spallation.

- **Evaporation pressure:** The evaporation pressure is another mechanism that can generate pulses having a high pressure at an exposed surface. The authors of [54,55] worked on the generation of a shock wave which was induced by a laser when an ambient gas struck the metal surface (see Figure 6.23). The idea is that a region of compressed air will exist between the moving vapour fumes coming from the metal surface and the surrounding (uncompressed) air. This will result in a shock front being induced in the air and ambient atmosphere. As the evaporating material has a high momentum, this mechanism would develop naturally without the need for the plasma to absorb laser energy. When this mechanism is modelled, high pressures are predicted.
- **Laser shock cleaning:** Cleaning can be accomplished by using a plasma shock wave (see Figure 6.24) [55] that is created when a laser pulse being intense in nature causes air to break down [56,57]. To avoid

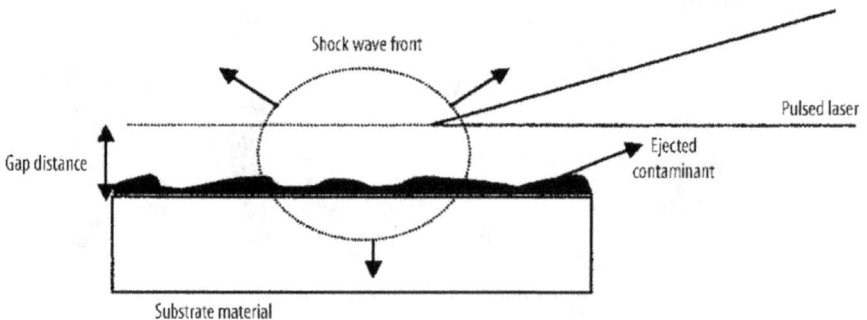

Figure 6.24 Laser shock cleaning [55].

direct laser encounter with the target material, the beam is pointed parallel to the surface and concentrated a few millimetres above the area that has to be cleaned. A shock wave is created which sounds like a finger snap as the gaseous contents start to break down and ionise. The usual shock front peak pressure for a plasma which is expanding and has the shape of a sphere in air is thought to be in the hundreds of MPa range. Since it is completely independent of the physical characteristics of the substrate and its contaminants, it does not run into the issues that other approaches do, where the effectiveness of removal depends on how well the substrate and the particles to be removed absorb light at a certain wavelength. Due to the incident beam not coming into direct contact with the workpiece, the danger of damage to the substrate is also reduced. In terms of the size of the area cleaned and the speed of cleaning, a higher cleaning efficiency is seen, much like with angular laser cleaning.

6.2.4.8.2 Applications

- Lasers can be used for cleaning, degreasing, depainting, derusting, and even for the removal of dirt which is ultrafine in size [58].
- If used on a large scale, they can strip paints as well [59].
- For Graffiti removal purposes.
- Laser cleaning could eventually replace water-intensive chemical wet techniques. Every day, a typical semiconductor manufacturing facility needs nearly 100,000 gallons of water. Finding a more cost-effective solution makes commercial sense given that the cost of disposing of 100,000 gallons of water which is contaminated with fluoride is approximately 450,000 dollars and the laser appears to provide precisely that.

6.2.4.8.3 Disadvantages of the Existing Technology

One of the greatest techniques for cleaning impurities from metals and non-metals is laser cleaning, although it isn't appropriate for all jobs. Larger surface areas may require more time to clean and may not yield the intended results because laser technology works well with small and targeted regions. The laser won't be able to clean something efficiently if you can't see it (for example, the inside region of a pipe). High cost is always a problem when using lasers as compared to other conventional techniques.

6.2.4.9 Medical Devices Manufactured Using Laser

- **Stents:** Stents and other types of micro-implants are expected to be in high demand both in developed and developing economies around the

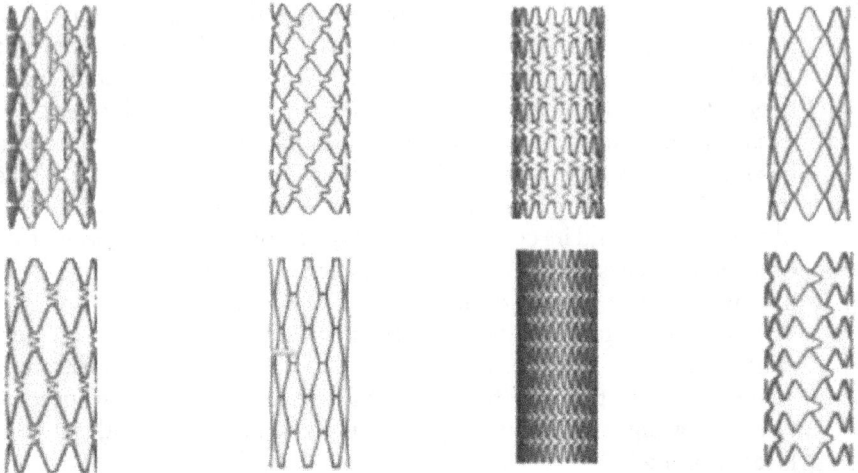

Figure 6.25 Different stent structures [62].

world. Stents are devices having an expandable grid made of a shape-memory alloy used to support blood arteries that are weak due to blockages and aneurysms (see Figure 6.25). A stent along with a balloon is put into the catheter that was used to remove the obstruction once it has been removed. The balloon is extended within the stent to stretch it and create a type of scaffold for the blood vessel, which is then inserted where it is needed as determined by an X-ray. The stent remains in place after the balloon and catheter have been taken out.

- **Laser marking:** Short pulses originating from a Nd:YAG laser can remove the top layer of the surface and make a mark on it thus modifying the morphology and characteristics of the surface. All medical equipment requires identification and traceability. The part number, size, compliance standard, company name, logo, rotation direction, and scale, if applicable, must all be included on the mark. The laser leaves a long-lasting, sterile mark. Also laser marking with high-frequency pulses minimises metallurgical damage.
- **Laser welding in the medical industry:** For implanted devices, it is crucial that the weld is durable and ductile, resistant to corrosion, biocompatible, aesthetically pleasant, and does not alter the part's size. Because of the high-quality welds and the regulated HAZ, the laser is widely employed in the medical industry. Some other applications include (see Figures 6.26 [60] and 6.27 [61]):
  - Pacemakers.
  - Welded catheter hubs.
  - Hearing aid devices.

Pacemaker

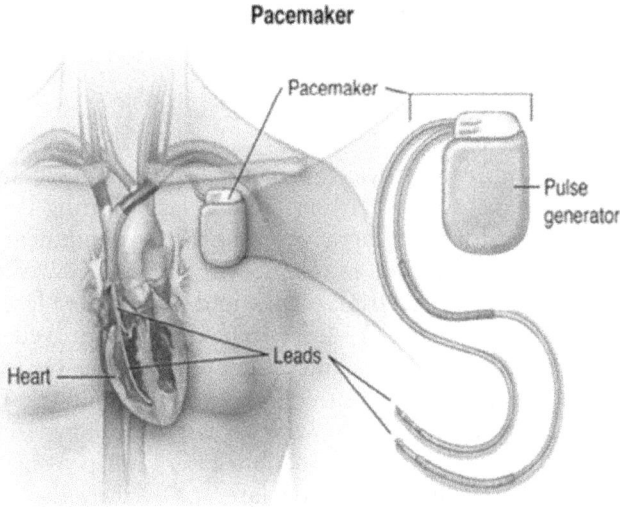

Figure 6.26 Pacemaker [60].

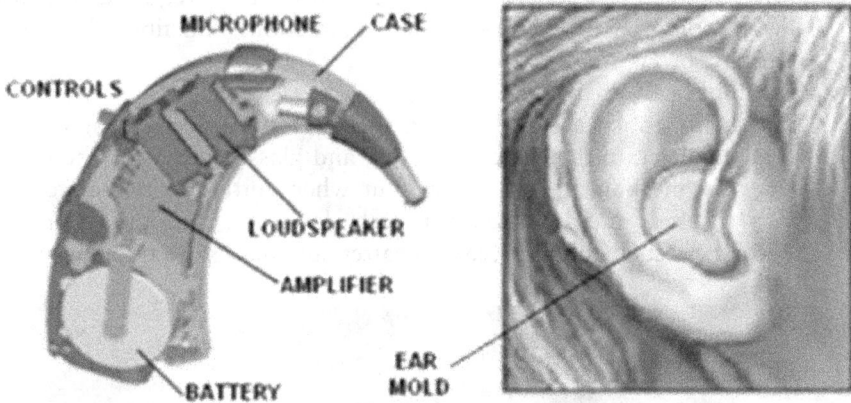

Figure 6.27 Hearing aid [61].

- **Wire stripping:** The fabrication of medical devices demands very precise procedures in addition to wiring systems that are ever increasingly intricate and sensitive. As smaller, more complex, and intricate medical devices continue to develop, laser wire stripping will remain the method of choice since it can handle many of the industry's issues. It is crucial to choose the right wires for the job and process them correctly when making medical devices. For a variety of medical applications, wire stripping using laser can provide superior quality processing.

6.2.5 Suitable Alternatives to Improve the Existing Technology

6.2.5.1 Laser Cutting

- Laser advancements

  High-power fibre and disc lasers are beginning to compete in the cutting business, even if carbon dioxide laser cutting is still a dominant contender for sheet metal profiling (as the equipment makers and their customers are well familiar with the procedures and cutting settings for various materials). Disk and high-power lasers have several advantages over CO_2 lasers, including improved energy efficiency, a more compact design, and more adaptable beam delivery via optical fibres. However, nothing is known about the fundamental traits of these lasers. If we want to cut metallic materials having thickness less than 5 mm, then fibre lasers are proven to be better performers than carbon dioxide lasers. On the other hand, if the thickness exceeds 5 mm, then carbon dioxide lasers perform better. In order for the laser manufacturers to provide fibre/disk laser cutting devices, they need the best cutting settings for a variety of materials at different thicknesses. Therefore, further research is required to completely comprehend beam–material interactions and fibre/disk laser cutting (see Figure 6.28).

- Technique advancements

  In terms of tool wear and cutting speed, laser cutting of fragile materials like ceramics, semiconductors, and glasses surpasses the challenges of mechanical machining. But when cutting brittle materials with a laser, micro-cracks, cut path deviation, and unfavourable thermal effects including recasts, spatter, and debris are frequently the

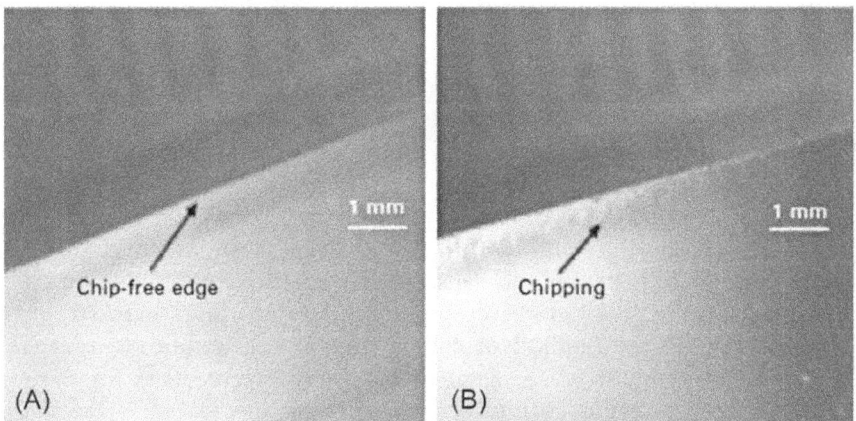

Figure 6.28 Cutting efficiency comparison of (A) Diode laser with (B) Mechanical cutting [63].

Figure 6.29 Using a continuous wave fibre laser to cut carbon composite, an illustration of a fibre pull out issue [64].

results. Different techniques have been employed for cutting of brittle materials like controlled fracture and thermal vaporisation techniques to name a few. By creating a single crack at temperatures much below the melting or phase transition temperatures, volume absorption has been shown to enable quick cutting of transparent and semi-transparent materials. Due to their low weight and suitable strengths, carbon or glass composite materials are being employed more frequently in the automotive, energy, aerospace, and sporting goods industries. The substantial material differences between the carbon or glass fibres and their polymer binders make laser cutting of composite materials and carbon challenging (see Figure 6.29). The polymer binder frequently dissolves when there is an excessive amount of heat transmitted through the fibres.

High-speed, brief pulse cutting can reduce the effect of heat, but the rate of material removal is not yet competitive on the market. Therefore, research is required to create laser cutting technology that is both faster and higher quality than mechanical and water jet cutting methods. Since the material attribute is anisotropic, modelling for this process is also difficult.

Materials with consistent thickness have so far been successfully sliced using laser technology. Cutting materials with changing thickness presents new difficulties. There are times when sheet materials delivered from several batches may not be precisely the same thickness. It is unclear how sensitive and important material thickness fluctuations are to cutting performance.

The laser cutting of steels with variable thickness has been the subject of some preliminary research, but more work is required to gain a fundamental understanding of how sensitive the laser cutting performance is to changes in material thickness while laser parameters are held constant. There are various applications where laser cutting of materials having a thickness, which is variable in nature, such as cutting a T-bar and pipes from one side, may be necessary. These applications include laser-based decommissioning of buildings and nuclear plants. Cut quality, however, might not be a crucial issue to take into account in certain applications.

6.2.5.2 Laser Welding

- **Laser advancements**
  Due to its unique benefits like the depth of penetration and superior gap tolerances, hybrid laser, and arc welding RnD activities have rapidly increased. It is more difficult than LBW because two distinct energies interact during the welding process. It has been discovered that the electrical arc can be stabilised by interacting a laser beam with it. The focus of current research is mostly on welding alloys and steels. Although the conventional techniques like arc and laser welding are well-known processes, hybrid laser arc welding is a relatively new development. Understanding the connections between process parameters, optimisation, and weld qualification, as well as process modelling, monitoring, and control systems, will be future difficulties.

  The depth of penetration for a laser welding process is about 25 mm but the rates for a carbon dioxide laser and a fibre laser are different. For a carbon dioxide laser, it is 1 mm/KW, and for a fibre laser, it is 2 mm/KW. In recent years, the depth of penetration has increased to upto 50 mm for steels, courtesy the fibre lasers. Thick-section welding is needed for several industries, like nuclear energy and shipbuilding, which is typically outside the scope of the lasers' current technology. There are still difficulties in creating laser welding methods that can have a direct competition with traditional methods for thick portions.
- **Technique advancements**
  - Although there has been extensive research into laser welding of different materials, many issues still need to be resolved. The development of intermetallics which are of brittle nature in the weld zones presents particular challenges.
  - Laser welding produces weld beads with parts above or below the actual surface of the material because of surface tension, rapid melt flow, and rapid solidification. Also possible is deformation of the weld sheets. This may not be an issue for the majority of

applications, and up to now, these areas have not been thoroughly researched. However, net shape welding is preferable for situations where fluid flow is influenced by surface topographical factors and for precise post-weld assembly.

Recent study at the University of Manchester has shown that mild steel sheets can be successfully net-welded. There are still difficulties in achieving net shape welding for various materials while retaining their properties like tensile and shear strengths.

6.2.5.3 Laser Drilling

- Laser advancements
  Lasers having an ultra-short pulse (like the femto and picosecond laser) can nowadays drill both metals and ceramic materials (see Figure 6.30). The average power for the picosecond laser is around 50 W whereas it is 400 W for a femtosecond laser. Due to the ultra-fast lasers' brief pulses, drilling at a high repetitive rate typically results in less debris, lower fluence, and significantly better surface polish than with longer pulse lasers. To fully comprehend the inter-actions between the beam and the material, however, a non-linear, multi-photon absorption is possible.
- Technique advancements
  There's always a problem of tool breakage with the conventional drilling techniques like the EDM or the mechanical drilling process, which makes it very difficult for such processes to drill very small holes, specifically at angles that are less than 90 degrees to the surface

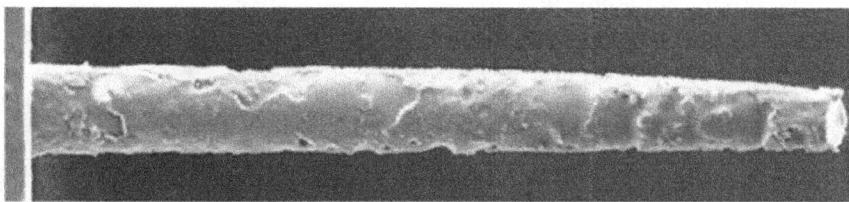

(A)

| 13x | 100 μm | EHT = 15.00 kV | Signal A = SE2 | Signal A = 1.000 | Date: Jul. 3, 2002 |
| Size = 1.180 μm | | WD = 17 mm | Signal B = inLens | | File name=WRO23e_imp_ |

(B)

Figure 6.30 (A) Laser-drilled hole. (B) Laser–EDM hybrid combination [65].

which is to be drilled. Combining laser technology with mechanical drilling or EDM would open up new possibilities for using both technologies' benefits. Figure 6.30 compares the drilling process done by a laser-drilled hole with a hybrid laser–EDM combination. When compared to EDM drilling, drilling time in this instance has been shortened, and breaking of the tool has greatly lowered. From this illustration, it is obvious that "hybrid" drilling would have advantages in producing holes quickly and with high quality.

6.2.5.4 Additive Manufacturing

The development of laser additive manufacturing (LAM) technologies opens up new avenues for the investigation and creation of aerospace materials. In terms of improving the laser head, changing the laser beam source, adding auxiliary energy fields, and combining with other processing techniques, various new techniques have steadily evolved [66]. Additional information on some of the important developing research goals will be provided in the section that follows:

- High-deposition-rate LDED technology also includes ultra-high-speed LDED technology. In order to protect large, high-quality parts from wear and corrosion, it tries to decrease layer thickness and enhance the surface deposition rate. The secret to using this technique is to increase the stand-off distance and place the melt-pool surface just below the focus of the powder material so that the powders are already melted when they arrive at the component surface [67].
- The laser wavelength for LAM processes is about 1064 nm and some alloys of copper and others have low absorptivity for such a wavelength. Either a blue or a green laser beam source having a wavelength around 515 nm has been created to address this problem. It has been demonstrated that pure copper and some copper alloys have a substantially better absorptivity for green or blue lasers and can produce parts without any flaws.
- Processing component segments with complicated structures are made possible by integrating LAM with conventional machine tools, which enables flexible switching between additive and subtractive production methods. Additionally, the LPBF process frequently generates tensile residual stress, which causes cracks to form and reduces mechanical performance. The creation of hybrid LAM is advantageous for dense, error-free aerospace parts that are designed and manufactured in a single piece.
- According to Mingareev and Richardson [68] (see Figure 6.31), here are some modifications that can be done in the LAM process:
  - Currently, process development typically begins with CAD data already created for conventional manufacturing processes, with

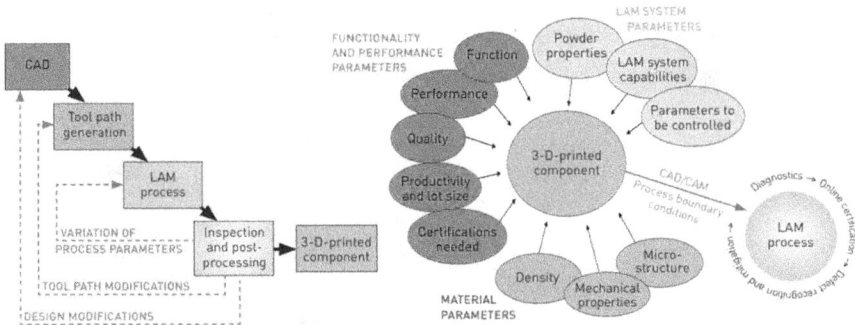

Figure 6.31 Comparison of the conventional approach and the one proposed by Mingareev and Richardson [68].

only minor tweaks and optimisations made specifically for LAM. The LAM tool route is then generated using empirical models, which primarily aim to reduce internal build faults. The process flow is hampered by equipment immaturity, the characteristics of the raw materials, melt pool instability, and varying solidification conditions.

- The initial phase is identifying the design's function and performance critical elements, as well as the part's topological restrictions (such as density and porosity requirements) and physical constraints (such as mechanical strength, stiffness, or heat conduction qualities). Critical raw material (powder) properties, a multi-scale framework of physical process conditions involved in laser–matrix interaction, dynamics of melting and solidification process, chemical transformations, and material microstructure evolution are all considered as part of the manufacturability process.

- According to this approach, only after a thorough, component-centred analysis are the real CAD data generated. A multi-dimensional data file (resolved in three dimensions of space and time) including the machine tool path, laser irradiation conditions, and other conditions customised for the component in question is the end result. We may take another look at how improvements in laser and photonic technology might contribute to the advancement of LAM with this integrated, "Component-Centred Design" [68] paradigm in mind. And in this situation, beam steering and shaping could be extremely helpful in enabling highly controllable spatial and temporal customisation of melt pools.

6.2.5.5 For Surface Treatments

Different micro or nano MMCs can be created and then coated on a metallic substrate using a laser beam and a sol-gel slurry. The fabrication of

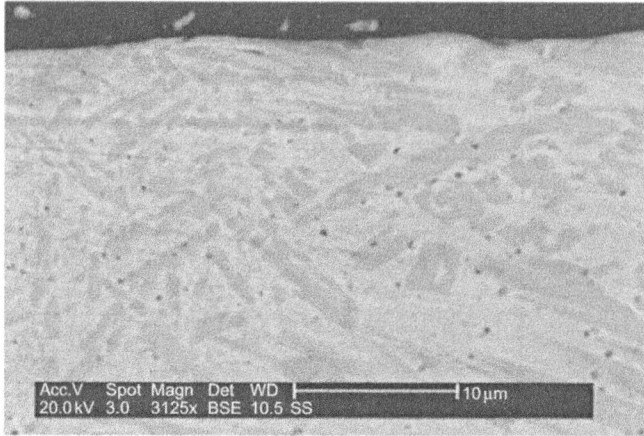

Figure 6.32 TiB, FeB, and Fe_2B particles embedded inside an iron matrix proposed by Choudhury et al. [69].

a surface MMC coating with either micro or nano TiB particles in iron by laser-sol-gel surface coating process is demonstrated as an example in Figure 6.32 in order to increase wear resistance. High-power diode, diode-pumped solid-state lasers, and also excimer lasers are employed more frequently for surface engineering due to their ability to deliver rectangular beams, even if CO_2 and Nd:YAG lasers are still frequently used. Nanomaterials and nanostructures are being used in more and more recent laser surface engineering studies.

6.3 SUMMARY, CONCLUSION, AND FUTURE CHALLENGES

In contemporary culture, laser materials processing is crucial. Despite the fact that laser processing has been used for a long period now (more than 40 years), there are still significant engineering, scientific, and technological obstacles to overcome on both the macro and nanoscales. Over the past two decades, the advancement of laser technology, including fibre, disc, diode, and ps/fs lasers, has been driven by industrial need for higher precision, higher speed, and improved energy efficiency. These novel laser sources present both new potential for the processing of materials and new difficulties in comprehending how they interact with materials. Research interests will continue to be drawn to the study of properly utilising these novel laser sources and discovering new capabilities and opportunities with these lasers.

In the context of laser cutting and drilling, the future challenges and developments include:

- **Lasers with high power**
  In order to achieve immediate high powers, superpulsing has produced several noteworthy results. A power that is too great could lead to explosive situations that lower the quality.
- **Additional sources of energy**
  The effects of traditional cutting, which employs oxygen as an extra energy source, have been observed. Arc augmentation requires considerable creativity to effectively locate the arc on the identical side as that of the laser and to solve the arc initiation issues. There are certain fields of study in composition of the gas being used. One of the more novel approaches is to blow Fe powder into an aluminium cut, so igniting a minor thermite reaction.
- **New and improved couplings**
  Currently used techniques include polarisation and coupling enhancements with high-power density. Radial polarisation has some potential. The behaviour of thin coatings on the melt pool front is poorly understood. There might be a specific gas mixture that would produce exactly the right thickness of the film for a certain substance.
- **Spot size**
  A smaller spot size is recommended. As a result, the melt zone would receive more energy and have a higher power density. This is accomplished using fibre lasers employing monomode fibre, however, the outcomes have been poor. This is because of the extremely small kerf widths that limit the melt's ability to flow out, but it's also possible that this is because the extremely high-power densities reached at these temperatures cause the oxidation reaction to stop and turn into a reducing reaction as predicted by the reaction's thermodynamics. Shorter wavelength lasers can also produce smaller spot sizes. A Gaussian high-powered UV beam's processing power hasn't been tested yet, but it should be amazing. However, there are a few safety issues that need to be addressed when employing shorter wavelengths.
- **Increase in fluidity**
  If the workpiece was vibrated ultrasonically, the operation was carried out in supergravity, or the melt's chemistry was altered to make it more fluid, the melt might be made to flow more easily. For instance, instead of oxygen, if we cut it with chlorine, it would either result in a vapour halide and no dross or a dross that was very fluidic in nature. Being denser than oxygen, chlorine gas would experience greater drag. An exothermic reaction is also involved. Therefore, chlorine ought to produce better results. But it can be lethal for the operator performing that particular job. All of these options do require consideration, but they appear challenging to an engineer.

Currently, there is a lot of interest in employing additive manufacturing technology in production to produce parts with almost identical designs and

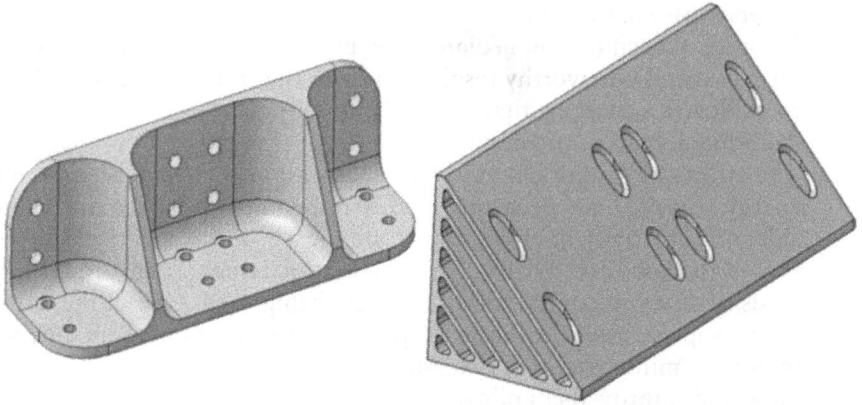

Figure 6.33 The part on the left is made through conventional CNC machine whereas it is made by AM on the right [70].

materials. And this is easily understood given that a specific design and material are picked following a careful study of their compatibility, and most frequently, these designs have proven to be the most suited. However, this makes it difficult for additive manufacturing to adapt. It is important to remember that the current design was either chosen because it suited the traditional manufacturing techniques at one point or because it was optimised for the conventional manufacturing process, such as forging or casting, when an existing component is being considered for processing via the additive route. Thus the best optimised design for an additive process like LENS or SLS may not be the one that results from this. For example, Vayre et al. [70] discussed the manufacturing of an angle bracket using both the conventional process and additive manufacturing as shown in Figure 6.33.

The materials employed in the procedure are the second factor for the future of AM. To process the exact materials that are now in use, the procedures are currently being commercialised and refined. But as time passes, more and more study is being done on the materials that additive manufacturing can produce. This does not imply that the existing materials will be entirely be discarded or that there will be a paradigm shift in the application-specific materials. It is crucial to recognise the role that additive manufacturing techniques play in creating specialised and unique components. As demonstrated by the example of creating a new human skull and forehead using LENS, it is able to create highly specialised and unique parts. Such specialised applications would also be possible with conventional manufacturing, but additive manufacturing will only be quicker and more accurate.

If we consider laser surface cleaning protection of cultural artefacts, metal surfaces, semiconductor industries, and optical lenses, all have benefited greatly from laser cleaning. In high-end sectors, it has received attention in the equipment field. As a result, the demands on the laser's performance are increased. Laser cleaning has a promising future, but still there are numerous obstacles to overcome which are addressed by Zhu et al. [71]:

- The technique for evaluating the results of laser cleaning for various items has not yet been developed. Each sort of filth has a unique density, composition, thickness, and level of adhesion, and each needs a specific set of cleaning process parameters to be removed successfully. Currently, it is challenging to quantify the temperature gradient and diffusion rate of energy produced by a laser due to the lack of information on the physical characteristics of various materials and the understanding of the pollution condition of compound composition. Additionally, the success of cleaning depends on precise laser parameter adjustment, and the best outcome must be ascertained using an interdisciplinary online monitoring technique.
- The specific equipment required for laser cleaning urgently needs a breakthrough because the cleaning effectiveness is too poor. Traditional fibre lasers are not very effective at cleaning attachments with intricate structures because of their low peak and pulse energies. The solid-state laser has a high single pulse energy, but its efficiency is generally low because of insufficient pulse repetition frequency.
- Modernising cleaning using intelligence and technology. Digital upgrading is an important approach to increase efficiency in addition to boosting the ability of laser cleaning equipment.
- Laser polishing, cleaning, and etching will overtake traditional manufacturing methods as average laser power increases and laser costs continue to fall. They have notable benefits in lowering consumable costs and are ideal for automation. Laser fine surface manufacturing technology can improve a variety of surface qualities of materials by performing micromachining which is non-destructive on the material surface. It is the future trend and direction of manufacturing development.
- Laser-based manufacturing is one of the technologies providing methodologies for sustainable manufacturing and significant versatility in terms of materials to be micro-machined as well as the structures that can be tailored [72–77]. Along with more advanced applications like drilling, patterning, welding, cutting, and laser marking, several new applications have arisen, from heavy vehicle to biomedical industry, that require more complicated operations like machining and manufacturing of sophisticated surface profiles in the bulk material [78–82]. These developments have promoted laser-based technologies as they provide benefits in terms of technology, production, and the economy [83–86].

REFERENCES

[1] H. Kaushal and G. Kaddoum, "Applications of lasers for tactical military operations,"*IEEE Access*, vol. 5, pp. 20736–20753, Sep. 2017. doi: 10.1109/ACCESS.2017.2755678.

[2] V. Coffey, "High-energy lasers: New advances in defense applications," *Optics and Photonics News*, vol. 25, no. 10, pp. 28–35, 2014.

[3] S. Affan Ahmed, M. Mohsin, and S. M. Zubair Ali, "Survey and technological analysis of laser and its defense applications," *Defence Technology*, vol. 17, no. 2, pp. 583–592, 2021, issn: 2214-9147. doi: 10.1016/ j.dt.2020.02.012. [Online]. Available: https://www.sciencedirect.com/science/article/pii/S22149 14719312231.

[4] R. Bogue, "Lasers in manufacturing: A review of technologies and applications," *Assembly Automation*, 2015.

[5] A. K. Dubey and V. Yadava, "Multi-objective optimisation of laser beam cutting process," *Optics & Laser Technology*, vol. 40, no. 3, pp. 562–570, 2008.

[6] G. Chryssolouris, *Laser machining: Theory and practice*. Springer Science & Business Media, 2013.

[7] P. Di Pietro and Y. Yao, "An investigation into characterizing and optimizing laser cutting quality—a review," *International Journal of Machine Tools and Manufacture*, vol. 34, no. 2, pp. 225–243, 1994, issn: 0890-6955. doi: 10.1016/0890-6955(94)90103-1. [Online]. Available: https://www.sciencedirect.com/science/article/pii/0890695594901031.

[8] G. Chryssolouris and P. Sheng, "Recent developments in three-dimensional laser machining," in *International Congress on Applications of Lasers & Electro-Optics*, Laser Institute of America, vol. 1990, 1990, pp. 281–293.

[9] W. M. Steen and J. Mazumder, *Laser material processing*. Springer Science & Business Media, 2010.

[10] D. Roessler, "New laser processing developments in the automotive industry," *The Industrial Laser Annual Handbook*, pp. 109–127, 1990.

[11] M. Warwick and M. Gordon, "Application studies using through-transmission laser welding of polymers," *Joining Plastics*, pp. 25–26, 2006.

[12] I. Jones and J. Rudlin, "Process monitoring methods in laser welding of plastics," in *Proceedings of the Conference Joining Plastics*, 2006.

[13] H. Y. Sarvestani, A. Akbarzadeh, H. Niknam, and K. Hermenean, "3D printed architected polymeric sandwich panels: Energy absorption and structural performance," *Composite Structures*, vol. 200, pp. 886–909, 2018.

[14] N. R. Mandal, *Ship construction and welding*. Springer, 2017, vol. 329.

[15] T. H. Maiman *et al.*, "Stimulated optical radiation in ruby," 1960.

[16] J. Coykendall, M. Cotteleer, J. Holdowsky, and M. Mahto, "3D opportunity in aerospace and defense: Additive manufacturing takes flight," *A Deloitte Series on Additive Manufacturing*, vol. 1, 2014.

[17] W. E. King, A. T. Anderson, R. M. Ferencz, *et al.*, "Laser powder bed fusion additive manufacturing of metals; physics, computational, and materials challenges," *Applied Physics Reviews*, vol. 2, no. 4, p. 041304, 2015.

[18] E. Malekipour and H. El-Mounayri, "Common defects and contributing parameters in powder bed fusion am process and their classification for online monitoring and control: A review," *The International Journal of Advanced Manufacturing Technology*, vol. 95, no. 1, pp. 527–550, 2018.

[19] H. Brooks, A. Rennie, T. Abram, J. McGovern, and F. Caron, "Variable fused deposition modelling: Analysis of benefits, concept design and tool path generation," in *Proceedings of the 5th International Conference on Advanced Research in Virtual and Rapid Prototyping*, Leiria, Portugal, 2011, pp. 511–517.

[20] H. Omar, N. I. M. Pauzi, M. Abu-Shariah, Z. M. Yusof, and S. B. Maail, "Microcracks pattern and the degree of weathering in granite," *Electronic Journal of Geotechnical Engineering*, vol. 14, 2009, pp. 1–21.

[21] W. J. Sames, F. List, S. Pannala, R. R. Dehoff, and S. S. Babu, "The metallurgy and processing science of metal additive manufacturing," *International Materials Reviews*, vol. 61, no. 5, pp. 315–360, 2016.

[22] L.-C. Zhang, H. Attar, M. Calin, and J. Eckert, "Review on manufacture by selective laser melting and properties of titanium based materials for biomedical applications," *Materials Technology*, vol. 31, no. 2, pp. 66–76, 2016.

[23] H. Gong, K. Rafi, H. Gu, T. Starr, and B. Stucker, "Analysis of defect generation in ti–6al–4v parts made using powder bed fusion additive manufacturing processes," *Additive Manufacturing*, vol. 1, pp. 87–98, 2014.

[24] C. Atwood, M. Griffith, L. Harwell, *et al.*, "Laser engineered net shaping (lens™): A tool for direct fabrication of metal parts," in *International Congress on Applications of Lasers & Electro-Optics*, Laser Institute of America, vol. 1998, 1998, E1–E7.

[25] W. Cong and F. Ning, "A fundamental investigation on ultrasonic vibration assisted laser engineered net shaping of stainless steel," *International Journal of Machine Tools and Manufacture*, vol. 121, pp. 61–69, 2017.

[26] B. Li, M. Zhang, Q. Lu, *et al.*, "Application and development of modern 3d printing technology in the field of orthopedics," *BioMed Research International*, vol. 2022, 2022, pp. 1–15. doi: 10.1155/2022/8759060.

[27] S. Husam, "Development of a selection program for additive manufacturing systems," Ph.D. dissertation, Stellenbosch: University of Stellenbosch, 2010.

[28] S. K. Saha, C. Divin, J. A. Cuadra, and R. M. Panas, "Effect of proximity of features on the damage threshold during submicron additive manufacturing via two-photon polymerization," *Journal of Micro and Nano-Manufacturing*, vol. 5, no. 3, 2017, p. 031002. doi: 10.1115/1.4036445.

[29] V. F. Paz, M. Emons, K. Obata, *et al.*, "Development of functional sub 100 nm structures with 3d two-photon polymerization technique and optical methods for characterization," *Journal of Laser Applications*, vol. 24, no. 4, p. 042004, 2012.

[30] J. Jiang, X. Xu, and J. Stringer, "Support structures for additive manufacturing: A review," *Journal of Manufacturing and Materials Processing*, vol. 2, no. 4, 2018, issn: 2504-4494. doi: 10.3390/jmmp2040064. [Online]. Available: https://www.mdpi.com/2504-4494/2/4/64.

[31] K. Shibata, "Recent trends in laser material processing in the Japanese automotive industry," *DGM Informationsgesellschaft mbH, Laser Treatment of Materials(Germany)*, pp. 433–442, 1992.

[32] L. Hector, "Focused energy beam work roll surface texturing science and technology," *Journal of Materials Processing and Manufacturing Science*, vol. 2, no. 1, p. 63, 1993.

[33] E. Johnston, G. Shannon, J. Spencer, and W. Steen, "Laser surface treatment of concrete," *ICALEO'97 Proceedings*, Orlando, A210–A218, 1997.

[34] J. M. Dowden, *The mathematics of thermal modeling: An introduction to the theory of laser material processing.* Chapman and Hall/CRC, 2001.

[35] E. Chen, K. Zhang, and J. Zou, "Laser cladding of a mg based mg–gd–y–zr alloy with al–si powders," *Applied Surface Science*, vol. 367, pp. 11–18, 2016.

[36] Y. Liu, Z. Guo, Y. Yang, *et al.*, "Laser (a pulsed Nd: YAG) cladding of AZ91D magnesium alloy with Al and Al2O3 powders," *Applied Surface Science*, vol. 253, no. 4, pp. 1722–1728, 2006.

[37] Q. Ma, X. Gao, and J. Li, "Microstructure performance and formation mechanism of laser alloying rare earth oxides modified nanocrystalline layer on ta7," *Physica E: Low-dimensional Systems and Nanostructures*, vol. 77, pp. 29–33, 2016.

[38] H. Zhang, Y. Pan, Y. He, and H. Jiao, "Microstructure and properties of 6FeNiCoSiCrAlTi high-entropy alloy coating prepared by laser cladding," *Applied Surface Science*, vol. 257, no. 6, pp. 2259–2263, 2011.

[39] X. Zeng, Z. Tao, B. Zhu, E. Zhou, and K. Cui, "Investigation of laser cladding ceramic-metal composite coatings: Processing modes and mechanisms," *Surface and Coatings Technology*, vol. 79, no. 1–3, pp. 209–217, 1996.

[40] Y. Li, P. Zhang, P. Bai, L. Wu, B. Liu, and Z. Zhao, "Microstructure and properties of Ti/TiBCN coating on 7075 aluminum alloy by laser cladding," *Surface and Coatings Technology*, vol. 334, pp. 142–149, 2018.

[41] R. Qin, X. Zhang, S. Guo, B. Sun, S. Tang, and W. Li, "Laser cladding of high Co–Ni secondary hardening steel on 18Cr2Ni4WA steel," *Surface and Coatings Technology*, vol. 285, pp. 242–248, 2016.

[42] T. M. Yue, H. Xie, X. Lin, H. Yang, and G. Meng, "Solidification behaviour in laser cladding of alcocrcufeni high-entropy alloy on magnesium substrates," *Journal of Alloys and Compounds*, vol. 587, pp. 588–593, 2014.

[43] A. A. Siddiqui, A. K. Dubey, and C. P. Paul, "A study of metallurgy and erosion in laser surface alloying of AlxCu0. 5FeNiTi high entropy alloy," *Surface and Coatings Technology*, vol. 361, pp. 27–34, 2019.

[44] A. A. Siddiqui, A. Dubey, and C. Paul, "Geometrical characteristics in laser surface alloying of a high-entropy alloy," *Lasers in Engineering*, vol. 43, pp. 237–259, 2019.

[45] A. A. Siddiqui and A. K. Dubey, "Laser surface treatment," *Engineering Steels and High Entropy-Alloys*, 2020.

[46] J. F. Asmus, C. G. Murphy, and W. H. Munk, "Studies on the interaction of laser radiation with art artifacts," *Proceeding of SPIE*, vol. 41, pp. 19–30, 1974.

[47] J. Asmus, *Laser consolidation tests international fund for monuments (final report)*, 1974.

[48] J. F. Asmus, Z. Mj, *et al.*, "Surface morphology of laser-cleaned stone," 1976.

[49] J. F. Asmus, "Light cleaning: Laser technology for surface preparation in the arts," *Technology and Conservation Boston*, vol. 3, no. 3, pp. 14–18, 1978.

[50] J. E. Asmus, "More light for art conservation," *IEEE Circuits and Devices Magazine*, vol. 2, no. 2, pp. 6–15, 1986.

[51] J. F. Asmus, "Light for art conservation," *Interdisciplinary Science Reviews*, vol. 12, no. 2, pp. 171–179, 1987.

[52] W. Zapka, W. Ziemlich, and A. C. Tam, "Efficient pulsed laser removal of 0.2 μm sized particles from a solid surface," *Applied Physics Letters*, vol. 58, no. 20, pp. 2217–2219, 1991.

[53] A. C. Tam, W. P. Leung, W. Zapka, and W. Ziemlich, "Laser-cleaning techniques for removal of surface particulates," *Journal of Applied Physics*, vol. 71, no. 7, pp. 3515–3523, 1992.

[54] M. Aden, E. Beyer, and G. Herziger, "Laser-induced vaporisation of metal as a Riemann problem," *Journal of Physics D: Applied Physics*, vol. 23, no. 6, p. 655, 1990.

[55] C. J. Knight, "Theoretical modeling of rapid surface vaporization with back pressure," *AIAA Journal*, vol. 17, no. 5, pp. 519–523, 1979.

[56] K. Watkins, C. Curran, and J.-M. Lee, "Two new mechanisms for laser cleaning using nd: Yag sources," *Journal of Cultural Heritage*, vol. 4, pp. 59–64, 2003.

[57] J. Lee and K. Watkins, "Removal of small particles on silicon wafer by laser induced airborne plasma shock waves," *Journal of Applied Physics*, vol. 89, no. 11, pp. 6496–6500, 2001.

[58] G. Daurelia, G. Chita, and M. Cinquepalmi, "New laser cleaning treatments: Cleaning, derusting, deoiling, depainting, deoxidising and de-greasing," in *Proceedings of the Conference Lasers and Optics in Manufacturing*, Munich, 1997, pp. 369–391.

[59] L. Ploner, "High power tea laser available for paint stripping," *Laser Focus World May*, pp. 26–27, 1995.

[60] J. Peal, I. Matthews, C. Runnett, H. Thomas, and D. Ripley, "An update on cardiac implantable electronic devices for the general physician," *Journal of the Royal College of Physicians of Edinburgh*, vol. 48, no. 2, pp. 141–147, 2018.

[61] L. Barbosa, R. Jordan, and J. Cordioli, "Structural dynamic analysis of a bte hearing aid case," in *7th National Congress of Mechanic Engineering—CONEM*, 2012.

[62] O. Kobo, M. Saada, S. R. Meisel, *et al.*, "Modern stents: Where are we going?" *Rambam Maimonides Medical Journal*, vol. 11, no. 2, 2020, p. e0017. doi: 10.5041/RMMJ.10403.

[63] S. Nisar, M. Sheikh, L. Li, and S. Safdar, "Effect of thermal stresses on chipfree diode laser cutting of glass," *Optics & Laser Technology*, vol. 41, no. 3, pp. 318–327, 2009.

[64] J. R. Lawrence, *Advances in laser materials processing: Technology, research and applications*. Woodhead Publishing, 2017.

[65] L. Li, C. Diver, J. Atkinson, R. Giedl-Wagner, and H. Helml, "Sequential laser and EDM micro-drilling for next generation fuel injection nozzle manufacture," *CIRP Annals*, vol. 55, no. 1, pp. 179–182, 2006.

[66] C. Tan, F. Weng, S. Sui, Y. Chew, and G. Bi, "Progress and perspectives in laser additive manufacturing of key aeroengine materials," *International Journal of Machine Tools and Manufacture*, vol. 170, p. 103804, 2021, issn: 0890-6955. doi: 10.1016/j.ijmachtools.2021.103804. [Online]. Available: https://www.sciencedirect.com/science/article/pii/S0890695521001139.

[67] T. Schopphoven, A. Gasser, K. Wissenbach, and R. Poprawe, "Investigations on ultra-high-speed laser material deposition as alternative for hard chrome plating and thermal spraying," *Journal of Laser Applications*, vol. 28, no. 2, p. 022501, 2016.

[68] I. Mingareev and M. Richardson, "Laser additive manufacturing: Going mainstream," *Optics and Photonics News*, vol. 28, no. 2, pp. 24–31, 2017.

[69] A. R. Choudhury, T. Ezz, and L. Li, "Synthesis of hard nano-structured metal matrix composite boride coatings using combined laser and sol–gel technology," *Materials Science and Engineering: A*, vol. 445, pp. 193–202, 2007.

[70] B. Vayre, F. Vignat, and F. Villeneuve, "Designing for additive manufacturing," *Procedia CIRP*, vol. 3, pp. 632–637, 2012, 45th CIRP Conference on Manufacturing Systems 2012, issn: 2212-8271. doi: 10.1016/j.procir.2012.07.108. [Online]. Available: https://www.sciencedirect.com/science/article/pii/S2212827112002806.

[71] G. Zhu, Z. Xu, Y. Jin, *et al.*, "Mechanism and application of laser cleaning: A review," *Optics and Lasers in Engineering*, vol. 157, p. 107130, 2022, issn: 0143-8166. doi: 10.1016/j.optlaseng.2022.107130. [Online]. Available: https://www.sciencedirect.com/science/article/pii/S0143816622001828.

[72] S. Bhoi, A. Prasad, A. Kumar, R. B. Sarkar, B. Mahto, C. S. Meena, and C. Pandey, "Experimental study to evaluate the wear performance of UHMWPE and XLPE material for orthopedics application," *Bioengineering*, vol. 9, p. 676, 2022. 10.3390/bioengineering9110676.

[73] V. P. Singh, S. Jain, A. Karn, A. Kumar, G. Dwivedi, C. S. Meena, and R. Cozzolino, "Mathematical modeling of efficiency evaluation of double-pass parallel flow solar air heater," *Sustainability*, vol. 14, p. 10535, 2022. 10.3390/su141710535.

[74] S. Bhoi, A. Kumar, A. Prasad, C. S. Meena, R. B. Sarkar, B. Mahto, and A. Ghosh, "Performance evaluation of different coating materials in delamination for micro-milling applications on high-speed steel substrate," *Micromachines*, vol. 13, p. 1277, 2022. 10.3390/mi13081277.

[75] A. K. S. Gangwar, P. S. Rao, and A. Kumar, "Bio-mechanical design and analysis of femur bone," *Materials Today: Proceedings*, vol. 44, Part 1, pp. 2179–2187, 2021, ISSN 2214-7853 10.1016/j.matpr.2020.12.282.

[76] A. Kumar, S. Gautam, Y. Gori, and P. Patil, "Dissimilar materials welded specimen analysis using FEA," *Journal of Critical Reviews*, vol. 6, no. 5, pp. 356–362, 2019, ISSN-2394-5125.

[77] A. Kumar and P. P. Patil, "Modal analysis of heavy vehicle truck transmission gearbox housing made from different materials," *Journal of Engineering Science and Technology*, vol. 11, no. 2, pp. 252–266, 2016.

[78] A. Kumar, H. Jaiswal, R. Jain, and P. P. Patil, "Free vibration and material mechanical properties influence based frequency and mode shape analysis of transmission gearbox casing," *Procedia Engineering*, vol. 97, pp. 1097–1106, 2014. DOI: 10.1016/j.proeng.2014.12.388.

[79] A. Kumar, S. I. Behmad, and P. P. Patil, "Vibration characterization and static analysis of cortical bone fracture based on finite element analysis," *Engineering and Automation Problems*, no 3, pp. 115–119, 2014. UDC- 621.

[80] P. P. Patil, S. C. Sharma, H. Jaiswal, and A. Kumar, "Modeling influence of tube material on vibration based EMMFS using ANFIS," *Procedia Materials Science*, vol. 6, pp. 1097–1103, 2014. DOI: 10.1016/j.mspro.2014.07.181.

[81] P. Patil, S. Sharma, A. Saini, and A. Kumar, "ANN modelling of Cu type omega vibration based mass flow sensor," *Procedia Technology*, vol. 14, pp. 260–265, 2014. DOI: 10.1016/j.protcy.2014.08.034.

[82] A. Prasad, G. Chakraborty, and A. Kumar, "Bio-based environmentally benign polymeric resorbable materials for orthopedic fixation applications," *Advanced Materials for Biomedical Applications*. CRC Press, Taylor & Francis, 2022, ISBN: 9781003344810. DOI: 10.1201/9781003344810-15.

[83] S. Singh, A. Kumar, S. K. Behura, and K. Verma, "Challenges and opportunities in nanomanufacturing," *Nanomanufacturing and Nanomaterials Design: Principles and Applications*. CRC Press, Taylor & Francis, 2022, ISBN: 9781003220602. DOI: 10.1201/9781003220602-2.

[84] S. Srivastava, D. Verma, S. Thusoo, A. Kumar, V. P. Singh, and R. Kumar, "Nanomanufacturing for energy conversion and storage devices," *Nanomanufacturing and Nanomaterials Design: Principles and Applications*. CRC Press, Taylor & Francis, 2022, ISBN: 9781003220602. DOI: 10.1201/9781003220602-10.

[85] G. Chakraborty, A. Prasad, and A. Kumar, "Processing of biodegradable composites," *Biodegradable Composites for Packaging Applications*. CRC Press, Taylor & Francis, 2022, ISBN: 978103227908. DOI: 10.1201/9781003227908-3.

[86] A. Kumar, Y. Gori, S. Rana, N. K. Sharma, and B. Yadav, "FEA of humerus bone fracture and healing," *Advanced Materials for Biomechanical Applications*. CRC Press (Taylor and Francis), 2022, ISBN: 9781032054490. DOI: 10.1201/9781003286806-14.

Chapter 7

Application of Laser-Based Manufacturing Processes for Aerospace Applications

Aditya Purohit, Brij Mohan Sharma, Tapas Bajpai, and Pankaj Kumar Gupta

CONTENTS

7.1 Introduction 157
7.2 LAM, a Viable Option for the Fabrication of Aerospace
    Components 159
    7.2.1 LAM processes 159
7.3 Laser Powder Bed Fusion 159
7.4 Laser-Directed Energy Deposition 160
7.5 Optimization of Parameters 160
7.6 Hybridization of the Process 162
7.7 Mitigation of Stresses 163
7.8 General Solutions to the Applications Through Findings 163
7.9 Conclusion 165
References 165

7.1 INTRODUCTION

Laser welding is increasingly used nowadays for aerospace applications. However, laser welding has many constraints, such as lower absorption of laser power in materials with a higher refractive index and reduced rate of weld pool formation because of lower viscous materials. The beam generated in LASER (Light Amplification by Stimulated emission of radiations) is monochromatic, coherent, and shows minimal divergence. The properties justify the use of lasers in welding, this is supported by the data that laser welding sales account for a total of 12% of the total laser applications [1]. The generation of photons in laser works on the principle, that inside an atom there are discrete energy levels. When an electron moving in the discrete level absorbs energy, it jumps into a higher energy shell, the electron being unstable in those shells comes back to its lower discrete shell, simultaneously releasing photons. This controlled emission of photons acts as the source of laser. Further, these photons are taken from the source to the target using optical fibers. This ensures that there is a safe distance

DOI: 10.1201/9781003402398-7

between the source and the target. The optical fibers also facilitate the use of 6-axis robots. There are two principles behind laser welding ie conduction limited and keyhole welding. In conduction limited only, solid–liquid and liquid–solid phase changes occur whereas in keyhole welding gaseous phase is also present. The industrial application in terms of aerospace manufacturing gives keyhole mode an advantage over conduction mode since it has an advantage in terms of welding of higher thickness. Due to its high penetration capabilities, it's widely used in the aerospace industry. Apart from the above characteristics and advantages, there are also various disadvantages such as the induction of residual stresses, distortions, lack of fluidity of filler material, problems associated with maintenance of welded components, low efficiency, etc. These problems were acknowledged by researchers across the globe, the problems were addressed by various researchers. This would be discussed in the following sections.

With years of advancement, additive manufacturing, specifically laser additive manufacturing (LAM) has become an efficient and economical technique to construct the aircraft components, structures, and assemblies. The characteristics required, i.e., light in weight, high strength, complex geometry, etc., for aerospace, making LAM the better technique to select among other techniques available. The general classification of LAM is shown in Figure 7.1.

SLS uses laser light to melt and fuse powders, which are then stacked one on top of the other to create printed parts based on sliced 3D model data, whereas in the SLM process, a high-density laser selectively scans metal powder kept at the bed which builds fully functional 3D parts by stacking the solidified layers of the powders [2]. According to a computer-aided design file, a highly concentrated laser beam is employed in the DMLS method to target a metal powder bed thus, metal particles are fused. This method is a development of the SLS procedure, which likewise creates 3D parts from fused metal powder layers [3]. LENS welds air-blown streams of

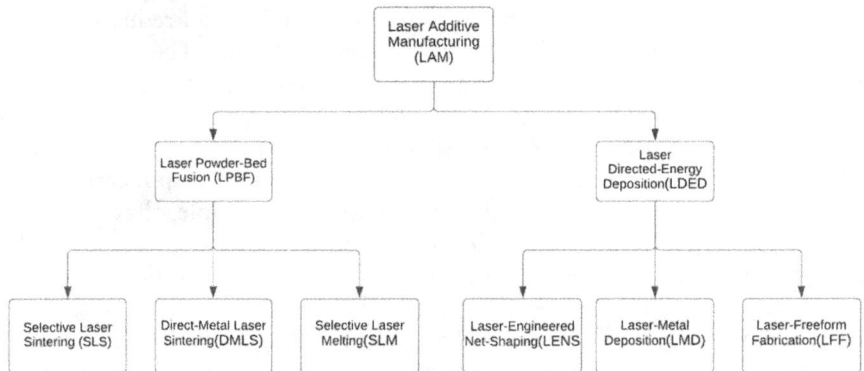

Figure 7.1 Classification of laser additive manufacturing.

metallic powders into 3D pieces over the course of hours. The process yields shapes that are sufficiently similar to the finished product [4]. LMD, the technique where most of the world's research seems to be focused, is performed by using a laser to melt a metal pool and then metal powder is deposited through a nozzle into the pool to create a layer and then another layer of material and so on till the part is not built completely whereas LFF, is used to create intricate 3D works of art by combining SLS and SLM [5].

7.2 LAM, A VIABLE OPTION FOR THE FABRICATION OF AEROSPACE COMPONENTS

The components such as fan blades, fan case, combustion chamber, high- and low-pressure compressor, turbine blades, etc., of the turbine of Boeing 787, are made of different materials [6]. Laser-based additive manufacturing is highly suitable for the production of such types of parts, components, and assemblies of aircraft because of the following properties:

- The complex geometry
- High buy-to-fly ratio
- High strength-to-weight ratio
- High customization with a small lot size
- Difficult to machine parts

7.2.1 LAM processes

The energy source for the LAM process is a laser beam. The focus of this section is to introduce two different LAM process types, namely LPBF and LDED. In LPBF, the powder is spread over the substrate, and in LDED, the powder is provided as a feedstock material. In ASTM Standard F2792-12a, these LAM classifications and the associated terminologies are described [7].

7.3 LASER POWDER BED FUSION

The concept of LPBF technology was first put up in 1996 by Meiners et al. from Germanise Fraunhofer ILT and Abe et al. from Japanese Osaka University [8]. In LPBF, a metal powder layer is first placed on the substrate by the recoater blade. In order to melt the metal powder, the laser beam is then employed to execute selective point-by-point scanning w.r.t. the 2D cross-sectional shape of the parts. When the laser beam recedes, these molten metal granules quickly solidify. Then, the platform will be pulled down by a predetermined depth equivalent to the layer thickness. The aforementioned procedure is then continued until the entire portion is created.

Table 7.1 List of aircraft components made using LAM

Name of Manufacturer	Product's Name	Product's Material	References
Aerojet Rocketdyne	Thrust-Chamber	Cu Alloy	[10]
Airbus	Reflector Bracket	Ti	[11,12]
	Bracket Connector of Cabin	Ti-6Al-4V	
Airbus & EOS	Aircraft Door Locking Shaft	Ti-6Al-4V	[13]
Airbus, EOS & Sogeti	Tail-plane bracket	Al-Si-10Mg	[14]
Amaero Engineering	Aero-Engine	Hastelloy	[11]
General Electric	Sump Cover	Co-Cr Alloy	[15–17]
	NACA inlet	Ti-6Al-4V	
	LEAP engine Fuel nozzle	Co-Cr Alloy	
NASA	Rocket-Injector	Inconel 625	[18,19]
	Pogo z-baffle	Inconel 718	
SpaceX	Components of Raptor Engine	Inconel Alloy	[20,20]
	Valve Body of Main Oxidizer	Not mentioned	

7.4 LASER-DIRECTED ENERGY DEPOSITION

In the early 1990s, LDED process was created by several research institutions throughout the world. Since the technical concepts are identical, the process is also named in various literature, such as laser solid forming (LSF), LMD, LENS, etc. While LPBF and LDED both discretize 3D models into 2D layers, LDED can employ feedstock material in both the forms wire or powder. Instead of being dispersed across a powder bed, the additive elements are introduced directly into the melt pool. Tan et al. [9] summarized that LDED can improve construction efficiency over LPBF methods by increasing power and size of laser beam. In addition, LDED is ideal for mending high-performance and high-value components, as well as producing gradient structures by the simultaneous feeding of various materials. However, LDED technique has limitations due to the difficulty in fabricating objects with exceedingly complicated geometry. A review of aircraft components made of different materials using LAM by various aircraft-making industries or R&D players such as Airbus, NASA, SpaceX is given in Table 7.1.

7.5 OPTIMIZATION OF PARAMETERS

Optimum parameter selection is necessary for the quality production by a process. Optimum parameter selection is associated with the optimization of the process. In order to optimize the process, Abioye et al. carried out their work on 5052- H32 Al sheets and suggested the optimum parameters for aerospace industry standards AWS D17.1 [21]. It was observed that industrial requirements were fulfilled by high energy and high power, as well as by low pulse energy and low peak power. Microstructural properties were studied and the relationship between microstructure and parameters was proposed. Optimization of energy is undoubtedly the need of the hour and the field of aerospace manufacturing is no different, Li et al. studied the energy parameters and optimized them for the energy consumption, with respect to weld quality [22]. CCD (Central composite design) was used to design the number of experiments, further kiring model was constructed between input and output. Finally, a tradeoff between parameters was obtained for low energy consumption and sound weld joint on Al 6061 alloy of 2-mm thickness.

Prior to carrying out actual LAM, optimization of process is necessary to create reliable processing conditions. Structures providing support to the part play a crucial role in the manufacturing of these parts using laser beams. Their primary duties include supporting overhanging structure, removing the heat produced, and limiting geometric distortions induced by internal stresses. By keeping this in mind, the Optimization of support structures for LAM of Ti6Al4V had been performed by Lindecke et al. [23]. It was suggested to choose the right design of the support to build the job using SLM by examining the properties of various types of structures used to support the part.

Process parameters also play the crucial role in LAM. Therefore, these process parameters have to be optimized. Parametric optimization of LAM of Inconel 625 was performed by Yang et al. using Taguchi and Grey relational analysis. In order to manufacture Inconel 625 samples using a LAM technique [24]. Researchers have analyzed the effects of key process parameters in LAM to manufacture Inconel 625 samples. The effect of laser power, scan speed of laser, feed rate of powder, and rate of overlapping as well as their interactions with each other was observed on width error and surface roughness in LAM. It was concluded that the overlapping rate has the highest impact on response parameters among all the processing factors. Li et al. [25] optimized SLM process parameters by using both the non-dominated sorting genetic algorithm-II (NSGA-II) and the ensemble of metamodels (EM) together. Among the process parameters namely, laser power, scan speed of laser, and layer thickness, layer thickness was having the maximum influence on the responses such as energy consumption, tensile strength, and surface roughness. Suzuki et al. [26] also investigated the optimum process conditions of the LPBF to fabricate tungsten carbide-cobalt (WC-Co) composites using the machine learning (Convolutional neural network) technique. It was reported that the laser power and laser spot size were having more influence than the laser scan speed to increase the WC decomposition region. Bonobo Optimizer algorithm was used by Singh et al. [27] to optimize process parameters of LMD to deposit WC and Co powder mixture on SS304 substrate for minimizing dilution. The optimized value of laser power, scan speed of laser, and feed rate of powder had been suggested as 500 W, 480 mm/min, and 560 mg/s, respectively, for the given materials. It was also shown that these optimized values were in good conformation with the result obtained experimentally. Therefore, optimization of process parameters is essential to build the part with desired quality using LAM processes.

7.6 HYBRIDIZATION OF THE PROCESS

Hybrid welding is one of the growing trends in welding. This form of welding introduces more than one kind of welding for fabrication. This utilizes the individual benefits of welding and tries to minimize the disadvantage associated with the welding, Zhou et al. used arc welding and laser welding in tandem in aerospace manufacturing. This technique produced superior-quality of joints with respect to bead geometry and microstructure characteristics [28]. The present research of the laser hybrid process is purely based on experimental work. Therefore, the researchers tried to involve the multiphysics of different techniques such as melting, solidification, convection, keyhole, and plasma formation in their experimentation.

Ola et al. comparatively analyzed the fuse weldability of aerospace grade alloys AA7075-T651 with respect to different weldings such as laser, hybrid laser, and cold wire. Microporosity was found to be one of the significant concerns of AA7075-T651 weldment, but the reduction of microstructure increased the susceptibility to Heat Affected Zone (HAZ) cracking [29]. The use of an argon-shielded environment coupled with cold wire welding in a controlled environment reduced the microstructure, with a reduction in HAZ cracking. A comparative analysis on the additively manufactured Ti-6Al-4V component using laser and arc beam deposition was done by Brandl et al. and it was concluded that the manufactured product had properties similar to aerospace material. The peculiar nature of the product was found in terms of direction, therefore the application of those products was restricted [30].

Hybridization of LAM process is needed as it is getting harder and harder to conform to the demands of integrated design of metal parts by using any LAM processes alone. Due to this, it has become a potential research area. Therefore, some researchers also proposed new hybrid LAM processes. Gong et al. [31] analyzed the microstructure and mechanical properties of Al-Si-10Mg alloy built by hybrid LAM processes (LPBF and DED). It was observed that there was good bonding between faces in the LPBF and DED zones and, the DED zone had small number of spherical pores. The LPBF zone has a greater average micro-hardness than the DED zone. It was concluded that components made of aluminum alloy can be produced using hybrid LAM (LPBF and DED) technology. A novel method, laser-shock modulation of molten-pool (LSMMP) of Fe-alloy is proposed by Liu et al. [32] This LSMMP method provides a new approach using a short-pulsed laser in hybrid LAM process. With the same pulsed laser energy input, it demonstrates more efficiency in residual stress control.

Ma et al. [33] compared the Laser-arc hybrid AM (LAHAM) of Cu-Cr-Zr alloy with wire arc additive manufacturing (WAAM). When compared to the WAAM sample, the LAHAM sample had finer grains and maximum texture indices and pole density intensities that were each lowered by 25.8% and 32.9%, respectively. AL-Cu alloy is widely utilized in aircraft due to its favorable property of high-strength-to-weight ratio. The LAM of Al-Cu alloy does have some drawbacks, such as cracks and weak strength. To overcome these drawbacks, a novel method Laser-Tungsten Inert Gas (TIG) additive manufacturing was proposed by Wu et al [34]. It was concluded that the mechanical properties such as elongation, ultimate tensile strength, and yield strength were higher than the samples which are fabricated by TIG, Cold Metal Transfer (CMT), and SLM methods. So in this way, researchers are putting their efforts to develop new hybrid methods to meet the challenges offered by advancement in terms of either novel materials or technology.

7.7 MITIGATION OF STRESSES

Since the aerospace application needs material with a high strength-to-weight ratio, to fulfill the above demand CFRP (Carbon Fiber Reinforced Polymer) came into picture. The dissimilar welding of different materials goes through different cooling cycles because of differences in material properties, because of differences in cooling cycles residual stresses are induced inside the material inadvertently. If these stresses are above yield strength then the induced residual stresses might lead to distortions or in some cases, failure. Wu et al. analyzed the welding of CFRP with SS and concluded that the introduction of residual stresses in some cases might lead to abrupt failure of the joint [35].

It's desirable to have compressive stresses in the weldment zone but on the contrary, the stresses obtained after the cooling cycle are tensile in nature. So, Li et al. studied the introduction of compressive stresses in Al alloys using laser penning in conjunction with FSW. The result showed that with the increase of laser penning and power density, the compressive residual stresses increased but after a limit asymptotic conditions occurred because of the hardening effect [36].

7.8 GENERAL SOLUTIONS TO THE APPLICATIONS THROUGH FINDINGS

Laser welding utilizes the potential of welding where conventional welding has reached its limit. As seen in various cases laser welding produces joints having the strength equal to that of the substrate material. Apart from that laser also gives the flexibility of creating hybrid structures such as metal–polymer, polymer–glass, polymer–wood, etc. Anwar et al. discussed the application of LTW between different materials and the feasibility of the process and concluded that LTW can only weld materials having chemical compatibility between them. The problem of welding transparent material was solved by the introduction of dies and other coloring pigments [37].

The effects of various parameters on the weld parameter were studied by various researchers. Magnesium alloys are suitable for aerospace applications because of their high strength-to-weight ratio. Zhang et al. worked on the alloy and found that by doing a modulation in the power of the laser source, they could improve the weld parameter, i.e., instead of using a constant power source if the sinusoidal source was used then melting volume and coupling energy efficiency increased, which were experimentally verified. Increasing the amplitude of the source also increased the equiaxial size of the grain. Since the intensity is directly proportional to A^2, thus ductility was induced [38].

Additive manufactured components are rigorously used in the aerospace and defence sector nowadays. With more usage arises new problems, such as maintenance of AM components. The maintenance of AM components is usually done by welding since the commonly used alloy for AM is Al alloy. The problem while repairing Al components is the introduction of porosity and softening of components. Peng et al. worked and concluded that the addition of Zr and Er as fillers in selective laser-melted AlSi10Mg resulted in a reduction of porosity of the AM components as well as the increase of UTS and YS compared to the as-welded conditions [39].

The offsetting of the laser beam while welding dissimilar material usually mitigates the effect of intermetallic formations. The offsetting is usually provided away from the material having a higher reflective index. Chen et al. studied the effect of processing parameters on the characteristic of Cu-SS joint using laser welding. It was observed that the weld mode transformed from brazing to fusion as the weld offset was shifted from SS to the interface of the base metals, respectively [40]. The effect of various parameters such as offset, oblique angle, welding speed, and power on mechanical behavior, microstructure, and appearance was studied. The optimal value of various input parameters, to get the desired output parameters were put forward by researchers. Chen et al., laser butt welded Ti-6Al-4V and Inconel 718 and concluded that brittle intermetallic phases could be reduced by deviating the laser beam on the Inconel side of the weld and crack-free welds of the alloy combination could be obtained by the use of higher power and velocity. Reduction in porosity was found when high velocity was employed because less time was made available for solidification, which thus promoted uniformity in the density of the structure [41].

The effect of offset on AZ31Mg and 60601 Al with Zn as a filler was investigated by Lv et al. [42]. The result concluded that with the increase in offset, the wettability of the filler material increased and the amount of intermetallic compound formed was kept within the acceptable range. With offset, the spread width of the filler material increased but the reaction depth decreased.

7.9 CONCLUSION

The paper discusses the rise of the application of modern lasers and additive manufacturing in the field of aerospace engineering. Various problems associated with laser welding in the aerospace industry were also discussed. The solutions opted for various problems such as optimization of parameters to increase the efficiency of the process and study the influence of various parameters (frequency, power, velocity, etc) on the output, hybridization of welding to improve porosity and microstructure, mitigation of residual stresses and miscellaneous (offsetting of the beam, effect of addition of various elements in filler, the effect of variation of amplitude, etc.) were reviewed. LAM technology is increasingly used in the aerospace field. The advancement of these manufacturing technologies such as hybrid LAM, hybrid laser welding, etc., encourages the liberation of structural design and development. Hence, the future technical sector of airplane production will strongly be impacted by promoting these laser-based additive manufacturing technologies.

REFERENCES

[1] N. Kashaev, V. Ventzke, and G. Çam, "Prospects of laser beam welding and friction stir welding processes for aluminum airframe structural applications," *J Manuf Process*, vol. 36, pp. 571–600, Dec. 2018, doi: 10.1016/J.JMAPRO.2018.10.005

[2] D. Gu, "Laser additive manufacturing (AM): Classification, processing philosophy, and metallurgical mechanisms," *Laser Additive Manufact High-Performance Mater*, pp. 15–71, 2015, doi: 10.1007/978-3-662-46089-4_2

[3] M. Shellabear and O. Nyrhilae, "DMLS – Development history and state of the art," Presented at LANE 2004 conference, Erlangen, Germany, Sept. 21–24, 2004.

[4] M. Izadi, A. Farzaneh, M. Mohammed, I. Gibson, and B. Rolfe, "A review of laser engineered net shaping (LENS) build and process parameters of metallic parts," *Rapid Prototyp J*, vol. 26, no. 6, pp. 1059–1078, June 2020, doi: 10.1108/RPJ-04-2018-0088/FULL/PDF

[5] M. Petersen and C. Emmelmann, "Development of new material systems for direct laser freeform fabrication," *International Congress Appl Lasers Electro-Optics*, vol. 2006, no. 1, p. 2002, Sep. 2018, doi: 10.2351/1.5060802

[6] A. P. Mouritz, "Introduction to aerospace materials," *Introduction to Aerospace Materials*, pp. 1–621, May 2012, doi: 10.2514/4.869198

[7] "Standard Terminology for Additive Manufacturing Technologies: ASTM F2792-12A - ASTM Committee F42 on Additive Manufacturing Technologies, American Society for Testing Materials, ASTM Committee F42 on Additive Manufacturing Technologies. Subcommittee F42.91 on Terminology - Google Books." https://books.google.co.in/books/about/Standard_Terminology_for_Additive_Manufa.html?id=H8ayrQEACAAJ&redir_esc=y (accessed Sep. 30, 2022).

[8] "Shaped body especially prototype or replacement part production," Dec. 1996. Granted: Feb 12, 1998 Applicants: Fraunhofer Ges Forschung Inventors: Meiners Wilhelm, Wissenbach Konrad Dr, Gasser Andres Dr.

[9] C. Tan, F. Weng, S. Sui, Y. Chew, and G. Bi, "Progress and perspectives in laser additive manufacturing of key aeroengine materials," *International Journal of Machine Tools and Manufacture*, vol. 170, Nov, p. 103804, 2021, doi: 10.1016/J.IJMACHTOOLS.2021.103804

[10] "Aerojet rocketdyne tests 3D printed thrust chamber with success - i3DMFG[TM]." https://www.i3dmfg.com/aerojet-rocketdyne-tests-3d-printed-thrust-chamber-success/ (accessed Sep. 30, 2022).

[11] "Success story aerospace facts additive manufacturing for the new A350 XWB", Accessed Sep. 30, 2022 [Online]. Available: www.sogeti-hightech.de.

[12] "Additive manufacturing enables 'bionic' aircraft designs – Make Parts Fast." https://www.makepartsfast.com/additive-manufacturing-enables-bionic-aircraft-designs/ (accessed Sep. 30, 2022).

[13] "3D Printing in Aviation I EOS GmbH." https://www.eos.info/en/all-3d-printing-applications/aerospace-3d-printing/airbus-case-study (accessed Sep. 30, 2022).

[14] "Manufacture of a small demonstrator aero-engine entirely through additive manufacturing (Aero-engine)". SIEF RP04-153: Aero-engine Final Report Summary 2017.

[15] "US Air Force and GE's collaboration on metal additive reaches first technology milestone with 3D printed sump cover for F110 engine I GE Additive." https://www.ge.com/additive/press-releases/us-air-force-and-ges-collaboration-metal-additive-reaches-first-technology-milestone (accessed Sep. 30, 2022).

[16] "Collaborate with the AddWorks [TM]* team at GE Additive to find a faster path to full-scale metal additive production. AddWorks from GE Additive". https://www.ge.com/additive/addworks.

[17] "New manufacturing milestone: 30,000 additive fuel nozzles I GE Additive." https://www.ge.com/additive/stories/new-manufacturing-milestone-30000-additive-fuel-nozzles (accessed Sep. 30, 2022).

[18] "Building a 3D Printed Rocket Engine." https://www.designworldonline.com/building-a-3d-printed-rocket-engine/ (accessed Sep. 30, 2022).

[19] "NASA Fires Up RS-25 Engines with 3D Printed Components to Highest Power Level – 3DPrint.com I The Voice of 3D Printing/Additive Manufacturing." https://3dprint.com/205157/nasa-tests-rs-25-engines/ (accessed Sep. 30, 2022).

[20] "SpaceX Reveals 3D-Printed Rocket Engine Parts I designnews.com." https://www.designnews.com/design-hardware-software/spacex-reveals-3d-printed-rocket-engine-parts (accessed Sep. 30, 2022).

[21] T. E. Abioye, H. Zuhailawati, S. Aizad, and A. S. Anasyida, "Geometrical, microstructural and mechanical characterization of pulse laser welded thin sheet 5052-H32 aluminium alloy for aerospace applications,"

Transact Nonferrous Metals Soc China, vol. 29, no. 4, pp. 667–679, Apr. 2019, doi: 10.1016/S1003-6326(19)64977-0

[22] Y. Li, M. Xiong, Y. He, J. Xiong, X. Tian, and P. Mativenga, "Multi-objective optimization of laser welding process parameters: The trade-offs between energy consumption and welding quality," *Opt Laser Technol*, vol. 149, p. 107861, May 2022, doi: 10.1016/J.OPTLASTEC.2022.107861

[23] P. N. J. Lindecke, H. Blunk, J. P. Wenzl, M. Möller, and C. Emmelmann, "Optimization of support structures for the laser additive manufacturing of TiAl6V4 parts," *Procedia CIRP*, vol. 74, pp. 53–58, Jan. 2018, doi: 10.101 6/J.PROCIR.2018.08.029

[24] B. Yang, Y. Lai, X. Yue, D. Wang, and Y. Zhao, "Parametric optimization of laser additive manufacturing of Inconel 625 using Taguchi method and grey relational analysis," *Scanning*, vol. 2020, 2020, doi: 10.1155/2020/9176509

[25] J. Li, J. Hu, L. Cao, S. Wang, H. Liu, and Q. Zhou, "Multi-objective process parameters optimization of SLM using the ensemble of metamodels," *Journal of Manufacturing Processes*, vol. 68, pp. 198–209, Aug. 2021, doi: 10.1016/J.JMAPRO.2021.05.038

[26] A. Suzuki, Y. Shiba, H. Ibe, N. Takata, and M. Kobashi, "Machine-learning assisted optimization of process parameters for controlling the microstructure in a laser powder bed fused WC/Co cemented carbide," *Addit Manuf*, vol. 59, p. 103089, Nov. 2022, doi: 10.1016/J.ADDMA.2022.103089

[27] A. K. Singh, et al., "Experimental investigation and parametric optimization for minimization of dilution during direct laser metal deposition of tungsten carbide and cobalt powder mixture on SS304 substrate," *Powder Technol*, vol. 390, pp. 339–353, Sep. 2021, doi: 10.1016/J.POWTEC.2021.05.056

[28] J. Zhou, T. T. Zhang, H. L. Tsai, and P. C. Wang, "Hybrid laser-arc welding in aerospace engineering," *Welding and Joining of Aerospace Materials*, pp. 123–156, Jan. 2021, doi: 10.1016/B978-0-12-819140-8.00005-5

[29] O. T. Ola and F. E. Doern, "Fusion weldability studies in aerospace AA7075-T651 using high-power continuous wave laser beam techniques," *Mater Des*, vol. 77, pp. 50–58, Jul. 2015, doi: 10.1016/J.MATDES.2015.03.064

[30] E. Brandl, B. Baufeld, C. Leyens, and R. Gault, "Additive manufactured Ti-6Al-4V using welding wire: Comparison of laser and arc beam deposition and evaluation with respect to aerospace material specifications," *Phys Procedia*, vol. 5, no. PART 2, pp. 595–606, Jan. 2010, doi: 10.1016/J.PHPRO.2010.08.087

[31] J. Gong, K. Wei, M. Liu, W. Song, X. Li, and X. Zeng, "Microstructure and mechanical properties of AlSi10Mg alloy built by laser powder bed fusion/ direct energy deposition hybrid laser additive manufacturing," *Addit Manuf*, vol. 59, p. 103160, Nov. 2022, doi: 10.1016/J.ADDMA.2022.103160

[32] J. Liu, S. Zhao, X. Zhang, X. Lin, and Y. Hu, "A laser-shock-enabled hybrid additive manufacturing strategy with molten pool modulation of Fe-based alloy," *J Manuf Process*, vol. 82, pp. 657–664, Oct. 2022, doi: 10.1016/ J.JMAPRO.2022.08.043

[33] G. Ma, et al., "Microstructure evaluation and resultant mechanical properties of laser-arc hybrid additive manufactured Cu-Cr-Zr alloy," *J Alloys Compd*, vol. 912, p. 165044, Aug. 2022, doi: 10.1016/J.JALLCOM.2022.165044

[34] D. Wu, et al., "Al–Cu alloy fabricated by novel laser-tungsten inert gas hybrid additive manufacturing," *Addit Manuf*, vol. 32, p. 100954, Mar. 2020, doi: 10.1016/J.ADDMA.2019.100954

[35] T. Wu, Y. Ma, H. Xia, P. Geng, T. Niendorf, and N. Ma, "Measurement and simulation of residual stresses in laser welded CFRP/steel lap joints," *Compos Struct*, vol. 292, p. 115687, Jul. 2022, doi: 10.1016/J.COMPSTRUCT.2022. 115687

[36] K. Li, X. He, L. Li, L. Yang, and J. Hu, "Residual stress distribution of aluminium-lithium alloy in hybrid process of friction stir welding and laser peening," *Opt Laser Technol*, vol. 152, p. 108149, Aug. 2022, doi: 10.1016/ J.OPTLASTEC.2022.108149

[37] G. Anwer and B. Acherjee, "Laser polymer welding process: Fundamentals and advancements," *Mater Today Proc*, vol. 61, pp. 34–42, Jan. 2022, doi: 10.1016/J.MATPR.2022.03.307

[38] M. Zhang, et al., "Impact of power modulation on weld appearance and mechanical properties during laser welding of AZ31B magnesium alloy," *Opt Laser Technol*, vol. 156, p. 108490, Dec. 2022, doi: 10.1016/ J.OPTLASTEC.2022.108490

[39] Z. Peng, et al., "Effect of Er and Zr addition on laser weldability of AlSi10Mg alloys fabricated by selective laser melting," *Mater Charact*, vol. 190, p. 112070, Aug. 2022, doi: 10.1016/J.MATCHAR.2022.112070

[40] H. C. Chen, A. J. Pinkerton, and L. Li, "Fibre laser welding of dissimilar alloys of Ti-6Al-4V and Inconel 718 for aerospace applications," *Int J Adv Manufact Technol*, vol. 52, no. 9–12, pp. 977–987, 2011, doi: 10.1007/s001 70-010-2791-3

[41] S. Chen, J. Huang, J. Xia, X. Zhao, and S. Lin, "Influence of processing parameters on the characteristics of stainless steel/copper laser welding," *Journal of Materials Processing Technology*, vol. 222. pp. 43–51, 2015. doi: 10.1016/j.jmatprotec.2015.03.003

[42] X. Lv and L. Liu, "Characteristics of laser-offset-TIG hybrid welding of AZ31Mg alloy with 6061Al alloy via Zn filler," *Opt Laser Technol*, vol. 152, p. 108126, Aug. 2022, doi: 10.1016/J.OPTLASTEC.2022.108126

Chapter 8

Laser Micromachining in Biomedical Industry

Avinash Kumar, Mohit Byadwal, Abhishek Kumar,
Ashwani Kumar, and Francis Luther King M

CONTENTS

8.1 Introduction 170
8.2 LMM and Texturing 173
    8.2.1 Laser Systems 174
    8.2.2 Applications in Biomedical Industry 176
8.3 Lasers and Materials LMM Devices 177
    8.3.1 Materials for LMM Devices 177
        8.3.1.1 Silicon 177
        8.3.1.2 Polymers 177
        8.3.1.3 Metals 178
        8.3.1.4 Ceramics 178
        8.3.1.5 Woods 178
        8.3.1.6 Paper 179
        8.3.1.7 Foams 179
        8.3.1.8 Textiles 179
    8.3.2 Common Industrial Lasers 181
        8.3.2.1 Carbon Dioxide Lasers 182
        8.3.2.2 Solid State Lasers 184
        8.3.2.3 Fiber Lasers 185
        8.3.2.4 Solid State Lasers in Medicines Today 185
    8.3.3 System Considerations 186
    8.3.4 Processing Considerations 187
8.4 Recent Trends in Biomedical Devices 187
    8.4.1 Bio-Implants 188
    8.4.2 Surgical Tools 189
    8.4.3 Lab on Chip 190
    8.4.4 Bio Sensors 191
    8.4.5 Prosthetics 192
    8.4.6 Angioplasty 192
    8.4.7 AI and Computer Software 194

DOI: 10.1201/9781003402398-8

8.5   Challenges in Biomedical Industry and Solution for Future.........195
8.6   Conclusion and Future Scope..198
References..199

8.1 INTRODUCTION

Micromachining is the application of specific procedures on micro and meso-size elements to produce components with high precision and extremely exact dimensional and geometric tolerances [1]. Because of the rapid advancement of Micro Electromechanical Systems (MEMS) research, industrial interest in small-scale manufacturing is growing. These are electrically powered mechanical devices having sizes in the micrometer range. The average MEMS device has a structure that is between 20 μm and 1 mm in size. It typically includes a data-processing central unit, a microprocessor, and numerous components that interact with the surroundings such as pressure sensors, accelerometers, or gyroscopes [2]. In the 1980s, MEMS manufacture was entirely reliant on procedures and materials acquired from integrated chip fabrication facilities. Molding, plating, wet and dry etching, and other methods capable of producing tiny devices were developed in the 1990s [3,117,120]. Many unusual materials were incorporated into MEMS devices to enable new uses. These materials have uses in medicine and biochemistry such as gas-permeable membranes, enzymes, biological cells, antibodies, and so on. MEMS have significant potential benefits over other classes of implanted devices due to their diminutive size, electrical nature, and capability to work quickly. In order to make the most of specific MEMS features, such as optical and electrical sensitivity or feature size similar to pertinent biological structures. The creation of retinal implants to correct blindness, neural implants to stimulate and record from the central nervous system, and very tiny needles for painless immunization are examples of uses. MEMS's digital capabilities might give drug delivery methods more temporal control than is now possible with polymer-based devices [4]. The capacity of MEMS to achieve precise control over the time of medication administration, along with their compact size, may prove to be of considerable use in the domain of actuating physiological systems within the human body. Hormones and growth factors are examples of molecules that are frequently highly powerful, and the timing of their administration has a significant impact on how they work. These powerful compounds have an impact on the body's signaling and regulatory systems, including the neurological and endocrine systems [4]. MEMS have the potential to actuate physiological systems due to their quick response times, physiological relevance of their actions, and the capacity to administer medications or an electrical stimulation from a device [5].

Laser micromachining (LMM) is one of the technologies providing methodologies for advanced manufacturing and significant versatility in terms of the materials to be machined as well as the structures that can be

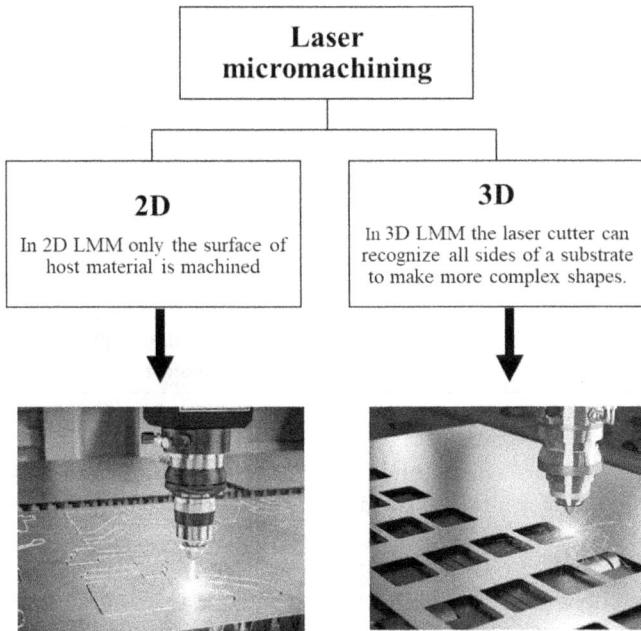

Figure 8.1 Showing two types of LMM.

tailored. Figure 8.1 shows two types of LMM. Unlike 2D laser cutting, which allows shapes to be obtained from flat sheet metal, in 3D laser cutting, 3D cutting is performed on parts already deformed by bending, pressing, or turning. In addition to more traditional applications such as drilling, thin-film patterning, laser marking, new applications have arisen, primarily in the fields of microfluidics and micro-optics, which require more complex operations like micromachining sophisticated surfaces into bulk material [6,116]. LMM has been developed to a stage now where more than one processing method is often suitable for any particular job. These developments have promoted laser-based technologies as they provide benefits in terms of technology, production, and the economy. There are two types of laser beam interactions with materials: thermal (rapidly raising the surface temperature till it melts) and ablative (rapidly increasing the surface temperature and causing mechanical shock on the substance) [7,118].

The following categories were used to categorize the micromachining processes: removal by mechanical force, melting vaporization, and others mentioned in Table 8.1 [8].

Material ablation process is divided into following stages:

- Deposited energy reaches a threshold value of ablation.
- The surface layer of a substance can evaporate by thermal (pyrolytic) or photolytic (UV radiation) means.

Table 8.1 Showing major methods grouped under micromachining process

Principle	Methods
Force	Cutting, grinding, USM
Melting vaporization	EBM, LBM
Ablation	LBM (excimer, short pulse)
Dissolution	ECM, Photo etching
Plastic deformation	Punching, press
Solidification	Molding, casting
Lamination	Stereolithography
Recomposition	Electroforming

- The produced plasma cloud is made up of reaction products, electrons and ions, and fragments of material particles.
- A plasma cloud absorbs and disperses the incoming pulsed laser energy.
- After reflection from the contact, the produced sound wave (into the material) may enhance the reaction products

Figure 8.2 shows different types of laser surface machining available for biomedical industries. A growing number of microsystems technology (MST) industries, including biomedicine, automobile production, telecommunications, display devices, printing technologies, and semiconductors, are using

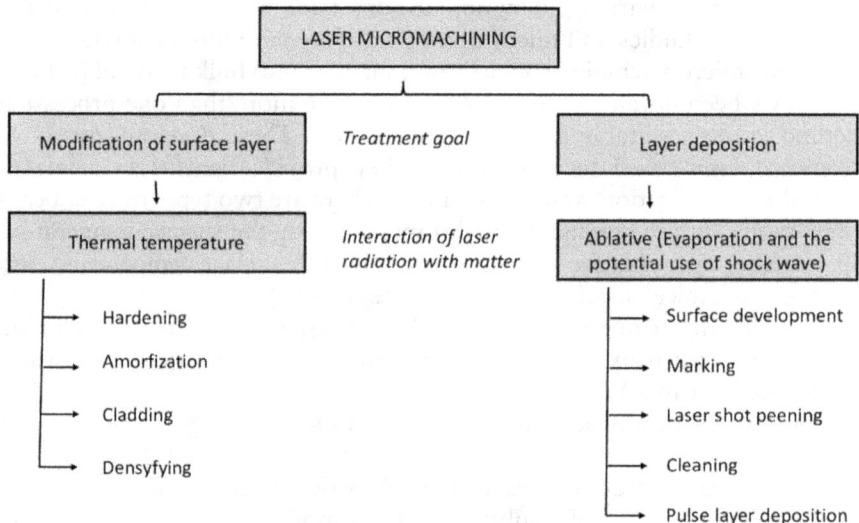

Figure 8.2 Types of laser surface micromachining.

laser systems [9]. These applications use lasers in a variety of ways, from early phases of research and development to full-scale manufacturing facilities. It is based on the process of laser ablation, in which material is degraded on interaction with laser energy. From early phases of research to full-scale manufacturing facilities, MST industries are using laser systems in several ways [9]. There have been several advancements and developments made to the laser systems and laser methodologies that are currently being employed as a result of the frequently incredibly stringent needs of the high specification items that are presently being considered [9].

Drug delivery methods are the focus of an intense research in order to overcome some of the drawbacks of usually adopted techniques, particularly when conventional approaches are sub-optimal in terms of efficiency, safety, and pain. A release of drugs, which has typically been accomplished by oral, pulmonary, transdermal, or injectable routes, frequently necessitates a high degree of precision to provide the best therapeutic effect [10]. The introduction of micromachined drug delivery systems enhances the therapeutic efficacy of pharmaceuticals. It has been stimulated by the development of micromachined actuators, valves, etc. employing biocompatible materials [11]. Microneedles, micropumps, and implantable drug-delivery microdevices have all been developed because of the growing interest in the field of drug-delivery microsystems and their immense potential [12]. A novel viewpoint on MEMS's capacity to activate diverse bodily systems like pacemakers and cardiology devices will also be discussed. The use of MEMS and micro-fabrication technology may enable greater physiological and logistical integration with the sensitive human body systems than is now possible in clinical or commercial therapies, such as for the actuation of the endocrine system [13]. The use of MEMS may be envisioned as a strong platform for delivering potent medicinal chemicals, whose timing is essential to their effectiveness and whose impact is magnified by the human body in a natural way [13].

8.2 LMM AND TEXTURING

LMM is a technique for machining and surface modification that makes use of a variety of wavelengths, wave shapes, and pulse durations. By utilizing ablation to provide big photons for absorption at shallow skin depths, the most precise bulk geometric change is accomplished. There are several lasers that can produce ultraviolet beams. One native ultraviolet laser that has a number of important advantages over rival sources is the excimer laser. To make use of these particular source features, many micromachining processes have been devised. The most intriguing of them can enable the machining of 2.5D structures in arrays across a substrate's surface.

Due to the special characteristics of micromachining lasers, surface patterns, and textures may be produced on a wide range of materials. High-resolution

Table 8.2 Machining processes compared to laser-machining processes

	Mechanical machining	Laser machining
Upfront cost?	V. High	High
Micromachining quality high or not?	High	High
Good yield or not?	Low	V. High
Any post-processing required?	Low	High
Machining fine features or not?	Low	V. High
Environmental conscious or not?	Low	High
Minimal noise or not?	Low	High
Rapid throughput or not?	V. High	High
Low cost of ownership or not?	High	High

forms with customized surface qualities may be patterned using novel laser methods. These distinctive surface textures and patterns can be applied to a wide range of biological applications. Table 8.2 shows the comparison and advantage of laser machining over other mechanical processes.

8.2.1 Laser Systems

The laser is the most crucial element of any LMM system. A suitable laser source needs to be selected after the LMM application has been determined [12]. The following laser qualities are often necessary for every LMM application: good beam quality, high pulse energy, high pulse repetition rates, short pulse duration, a small concentrated spot size, dependability, and ease of operation with easily configurable system settings. Due to the high initial and ongoing costs of industrial lasers, the LMM application and the material being processed must be carefully considered [14]. Although a laser source can have many different characteristics, wavelength, the output power, temporal and spatial modes, and output beam width are some of the most crucial ones when choosing a laser. The most essential component of the laser and the main cost driver is output power. The laser source should be designed into the LMM system with the suitable output power for the workpiece material [14]. Longer processing times or the impossibility to manufacture the needed material will come from selecting a laser system with insufficient power; on the other hand, selecting a laser system with excessive power will increase the cost of the entire system [15]. Since LMM is still a relatively new field, the designer's experience is used to estimate the laser power requirements. By examining a specific LMM process model and comparing material attributes with laser operating parameters, an approximate estimation of the laser power needs may be generated [16].

There are two ways that lasers can work: continuous wave mode and pulsed mode. Continuous wave operation provides the substance with energy without any interruption. In pulsed mode, energy is given in quick pulses (bursts). There are benefits and drawbacks to both of these modes. Despite having a low average power, pulsed lasers can create pulses with high instantaneous energies. Pulsed lasers 29 can handle materials that are challenging to process, but they cannot guarantee a high-grade surface quality [17]. The use of pulsed lasers with short pulse durations and very high repetition rates is frequently required for LMM applications because they produce surfaces with a high degree of quality. Continuous wave laser processing, however, can result in excessive heating (a large heat-affected zone) and the removal of nonirradiated parts of the workpiece material. Another important consideration when choosing a laser is the spatial mode [18]. Transverse electromagnetic mode is the factor that determines the beam profile (TEM). The laser beam mode or profile is the term used to describe the intensity (irradiance or power per unit area) distribution in a plane perpendicular to the beam propagation axis. The beam mode describes the distribution of energy intensity across the beam cross-section. As you move away from the laser's output mirror, the profile changes [18].

For LMM applications, the basic mode, which has a Gaussian spatial distribution, is frequently regarded as the best (most of the energy is concentrated in the center of the beam) [19]. A spot with a smaller width can focus a laser beam with a nearly Gaussian distribution. The ratio of a laser beam's multimode diameter-divergence product to its fundamental beam diameter-divergence product is known as the beam quality factor (M2). The spatial distribution approaches Gaussian behavior when this parameter's minimal value = 1, is reached, indicating that the laser beam can focus more effectively. If we compare different laser kinds, fiber lasers have the lowest beam quality factor. The wavelength of the laser has a significant impact on the LMM process, owing to the laser and material interaction [20].

Since materials are handled using a relatively cool ablation process, shorter wavelength lasers are more suitable for LMM. In other words, to remove material at shorter wavelengths, photons with higher energy break atomic and molecular bonds. In contrast, longer-wavelength lasers heat and boil materials locally, which is how they treat materials thermally. Smaller spot sizes can be produced by focusing shorter wavelength lasers. Nd:YAG solid-state lasers (1.06 or 0.355 μm wavelength), pulsed or continuous wave CO_2 gas lasers (10.6 μm wavelength), and excimer lasers with nanosecond pulses are the laser types that are most frequently used for LMM applications (248 or 193 nm wavelength). All of the aforementioned lasers are used in pulsed mode to achieve the maximum machining quality. CO_2 lasers can only focus a small amount of energy due to their long wavelength, making them unsuitable for use in applications requiring very small feature sizes [21]. Excimer lasers are the most economical lasers; however, they have a low repetition rate and subpar beam quality [22]. Although mask projection is the process that excimer lasers are used for most commonly, these shortcomings significantly affect how well they work when used for direct writing techniques. The pulse lengths of the lasers mentioned above are what set them apart. Pico-second and femtosecond lasers, which are employed in a range of micromachining applications, are the result of recent developments in laser technology. Because they frequently have extraordinarily high peak pulse powers, ultrashort pulse duration lasers can machine most substrates with little reliance on absorption effects. Ultra-short pulsed lasers prevent secondary effects of LMM such as thermal damage, recasting, and debris because they do not generate heat in the target material. These lasers continue to be quite expensive [21].

8.2.2 Applications in Biomedical Industry

There are lots of products/devices in biomedical industry, which includes LMM. Stents, intraocular lenses, prostheses, and catheters are just a few examples of the increasingly complicated implantable medical devices. These devices' feature sizes are getting smaller to accommodate new applications and enhance patient outcomes. Metal stents, including drug-eluting metal stents, can be deployed in smaller coronary, peripheral, and neurovascular blood arteries by having their size reduced essential areas that will be treated. According to several research, the amount of metal used inside the vessel and clinical outcomes are related. As a result, stent manufacturers have been inspired to develop tubes with thinner walls, smaller diameters, and more intricate stent characteristics. According to a recent report, the next generation of coronary stents would include struts and linkages among other features that are nearly half the size of current stent features. Another approach is to give stents and prosthesis a regulated surface texture or geometry to increase biocompatibility, such as to lower the danger of restenosis (the recurrence of arterial narrowing after treatment). The fabrication process is finding it difficult to incorporate new materials and reduce feature size. As stent strut size decrease, not only does the cut edge quality become more important, but it also gets harder to produce. For smaller stents, the heat-affected zone (HAZ), or damage from heating, starts to function as a limiting factor, making it challenging to remove damaged material through post-processing. Fortunately, improvements in LMM are moving us closer to solving these problems.

In manufacturing of medical balloons, lasers play a critical role in creating detailed, high-precision medical balloon features that no other methods can duplicate including texturing, grooving, and section thinning. With the latest laser technologies, tools, and materials, medical balloons can be manufactured with microscale features, from a large variety of materials, for increasingly innovative neurovascular, cardiovascular, gastrointestinal, urological, and catheterization procedures [23,121] (Figure 8.3).

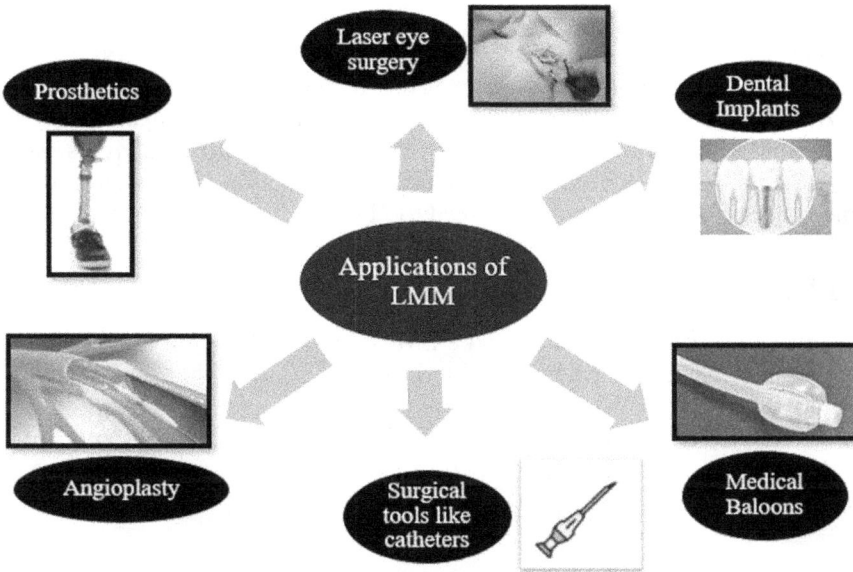

Figure 8.3 Various applications of LMM in biomedical industry.

8.3 LASERS AND MATERIALS LMM DEVICES

8.3.1 Materials for LMM Devices

8.3.1.1 Silicon

Silicon (Si) is the material used to make the majority of integrated circuits used in current consumer devices. It is also an appealing material for the manufacture of LMM due to its numerous beneficial chemical and mechanical properties: Single crystalline Si is a nearly perfect Hookean material. This means that there is nearly little hysteresis and hence almost no energy loss when Si is bent. This feature makes Si the best material for applications requiring many tiny movements and excellent dependability, since Si exhibits very low degradation and may attain lifespans in the billions to trillions of cycles. Si laser micro cutting and laser micro hole drilling are well suited for precise and chipping-free cutting of very thin as well as extremely thick Si material [21].

8.3.1.2 Polymers

Polymers are ideal for laser processing; they respond extremely well in the ultraviolet range of the light spectrum. Based on photodecomposition of the material, LMM is clean, non-contact, and causes no HAZs or material degradation. Different types of plastics display superior cutting quality at a specific wavelength over another; it is a matter of assessing each application and developing a working plan for the best possible outcome. Even though the electronics sector offers economies of scale to the Si industry, crystalline Si remains a complicated and somewhat costly material to produce. Whereas polymers may be made in large quantities and with a wide range of material properties [19]. LMM devices, which can be made from polymers using processes such as injection molding, embossing, are particularly well suited to micro-fluidic uses such as disposable blood testing cartridges [22].

8.3.1.3 Metals

Metals can also be utilized to make LMM components. While metals may not have some of the mechanical qualities and benefits that Si does, when employed within their constraints, metals can demonstrate very high levels of dependability. Metals can be deposited using electroplating, evaporation, and sputtering. Gold, nickel, copper, chromium, platinum, and silver are all commonly utilized metals [20].

8.3.1.4 Ceramics

Because of beneficial material property combinations, Si, and titanium nitrides, as well as Si carbide and other ceramics, are increasingly used in LMM fabrication. Lasers have more recently been employed to machine ceramics. Hard, brittle, and abrasive materials may be processed by lasers [24]. As a non-contact method, laser machining does away with issues with tool wear, tool deflections and vibrations, and cutting forces on the workpiece. Additionally, lasers' performance is unaffected by the mechanical qualities of the material and can be simply automated. They can also be easily converted to flexible manufacturing systems. Ceramics and other materials have been processed using all three of the main types of lasers: carbon dioxide, ND:YAG (neodymium yttrium aluminum garnet), and excimer laser.

8.3.1.5 Woods

For those who are unfamiliar with the method, wood makes the greatest laser-cutting material. All types of wood can be sliced using laser technology. However, it's also crucial to consider the thickness.

Plywood, bamboo, and MDF are common types of wood used in laser cutting (medium-density fiberboard). They are perfect because they are affordable and their thickness may be adjusted. They are therefore perfect for testing out ideas before investing in pricey wood materials. Due to its adaptability, wood is a widely utilized material in a variety of industries. Effective CO_2 laser technology is being employed more and more. The material is more suited for laser cutting and engraving the more homogenous the wooden structures are. Due to the thermal laser process, additional characteristics such as moisture content, density, oil, and resin content have an impact on the laser cut's quality and speed. There will be a weak to strong oxidizing cut edge when processing wood materials with a laser. This means it is impossible to prevent a darkening. These frequently vary greatly on the type of wood used as well as its water and resin content.

8.3.1.6 Paper

Many paper-based laser-cut products will be recognizable to you. Those invites with cutting-edge designs are a common illustration. They are less expensive and simpler to cut than most laser-cutting materials. They are recyclable and biodegradable, making them environmentally friendly as well. A superior paper processing technique is laser cutting. High-precision contours are made feasible by the laser, and the material is not mechanically stressed. During laser cutting, paper does not burn; instead, it suddenly evaporates. As a result, the cutting edges only experience minimal temperature loads.

8.3.1.7 Foams

Foams are used to create household goods and packaging. They are a fantastic alternative to check out even if they might not be as well-known as other laser-cutting materials. Popular foam products like Depron foam, EPM, Gator foam, and Polyester are available for application (PES). In the past few years, foams' industrial applications have grown significantly. There are many different materials available on the foam market now for a variety of applications. The industry is increasingly learning how to use a laser beam as a tool for cutting foam. In some situations, the laser can be a very effective substitute for traditional machining techniques.

8.3.1.8 Textiles

New materials for laser cutting include textiles. Digital embroidery can be done on textiles with laser cutting. The operation can be used in various textile types, including leather, nylon, polyester, silk, etc. Textile cutting is best suited for large-format laser systems. Lasers are being used to cut contemporary fabrics, spacer fabrics, needle felt, glass fiber fabric, thermoresistant textiles, and other textiles. Important arguments in favor of the employment of laser technology in the processing of textiles include, in particular, the contactless processing with laser beam, the distortion-less cutting involved, with very high precision. Both natural and synthetic materials can be used to make textiles. Fabrics, knitted fabrics, non-woven fabrics, and felts are typical textiles for laser cutting. Their primary constituents are fibers that have a high degree of suppleness and flexibility.

Table 8.3 shows a comparison of available materials for LMM. It compares different materials for biomedical industry based on the optimal

Table 8.3 Comparison of materials for Biomedical applications

	Acrylic	Textiles	Polyester	Polyamide	Foams
Optimal cutting method	Carbon dioxide laser	Carbon dioxide laser	Carbon dioxide laser	Carbon dioxide laser	Solid state laser
Advantages of processing by LMM	(1) Less waste produced due to contactless laser cutting. (2) No need to clamp and fix of acrylic sheets.	(1) No distortion of fabric due to contactless processing. (2) Precise cutting of very filigree details.	(1) No material fixation required due to vacuum table. (2) No tool wear and thus, a constantly high cutting quality.	(1) Clean and perfect cut edges no post-processing necessary. (2) Flexibility in contour choice no tool preparation or tool wear.	(1) Engravings or markings in one operation. (2) No distortion of material during the cutting process.
Applications in biomedical industry	Signage and labeling of medicines and packages.	Protective masks – laser cutting of textiles for the mouth-nose protection.	Used in manufacturing of different medical devices, like sutures, plates, bone fixation devices, stents, etc.	Used to manufacture catheters, and dentures due to its biocompatibility property.	Biomedical foams are used for Tissue Engineering Applications.

cutting method and advantages of processing by LMM. It also briefly explains the field in biomedical industry, where these LMM can be applied.

8.3.2 Common Industrial Lasers

For the production of medical devices, lasers provide distinctive processing capabilities. Although competing technologies may provide equal technological capabilities, the utilization of LMM technology enables more cost-effective solutions compared to other current approaches [25]. In many other cases, laser-based manufacturing could be the sole technological means of resolving certain fabrication issues. It becomes crucial that the device produced must take into account both tangible factors such as processing velocity, feature size, and budget as well as intangible ones, such as "feature quality" and "customer friendliness". It suggests the most practical, efficient strategy [26]. Table 8.4 shows a comparison between applications and materials based on laser wavelength.

Table 8.4 Comparison between applications and materials based on laser wavelength [27,28,123]

	Host material	Applications	Wavelength
Solid state laser	A "dopant" such as neodymium, chromium, erbium, thulium, or ytterbium is introduced to a glass or crystalline substance that serves as the host material.	Red-green-blue light sources in laser printers and projectors, medical uses such as eye surgery, and metal processing	700–1000 nm
Fiber laser	Metal sheets (mainly stainless steel)	Removing rust, paint, oxides, and other contaminants cutting metal sheets	840–2200 nm
CO_2 laser	Non-metallic materials such as plastics but can also cut soft metals	Carbon dioxide laser surgery carbon dioxide laser cutting and welding	10,000 nm and above
Liquid lasers	Ytterbium-doped glass, neodymium-doped yttrium aluminium garnet, neodymium-doped glass and sapphire	They are used in laser medicine, spectroscopy, birthmark removal, and isotope separation	50–100 nm
Semiconductor lasers	Neodymium-doped yttrium aluminum garnet	Ablation of tissues during surgical treatment, optical diagnostics of tissues, tissue regeneration, tissue bonding, etc.	600–1100 nm

In industrial micromachining applications, the three most prevalent laser types are excimer, solid-state (primarily Nd:YAG and Nd:YLF), and CO_2 lasers. Each has distinctive qualities of its own, and when combined, they offer a range of characteristics that are quite complimentary [26].

8.3.2.1 Carbon Dioxide Lasers

The most well-known and often utilized photon sources are CO_2 lasers. Because of their widespread acceptance, simplicity, and convenience of use, they have been used in industry since the 1960s and are still widely used today. Although some of these lasers come in multi-kilowatt power variations, the majority of medical micromachining uses less than 150 W on average. CO_2 lasers are infrared photon sources because the wavelength of the photons they release is around 10 μm. The ultimate feature resolution varies depending on the wavelength employed, but for the fundamental frequencies, it is around 2 μm, while the ultimate practical resolution ranges from 10 to 25 μm. The primary energy levels of the molecules' rotational vibrations govern the underlying physics of material interaction with infrared photons. The majority of materials absorb photons with a wavelength of 10 nm, which raises the temperature of the surrounding environment by increasing the frequency of molecules vibrating. As a result, thermal contact is the main interaction. These thermal effects are noticeable at the microscopic level in the majority of materials. The CO_2 laser offers a quick way to treat materials if these side effects are tolerable. The CO_2 laser can potentially produce a 10 μm feature resolution, however, it is often only capable of 50 μm resolution since the maximum possible feature resolution depends on the photons' wavelength [29].

Finally, the CO_2 laser is configured for focal-point machining in most (but not all) applications, which implies that only one feature may typically be machined at a time. According to a study of the most recent literature, FP provides considerable advantages over traditional ablative laser resurfacing procedures in terms of lesser recovery times and side effect profiles that are neutral (CO_2). Though traditional ablative laser resurfacing produced skin tightening results comparable to surgical correction, the long-term risks of prolonged dispigmentation and potential scarring, as well as the side effects including several months of swelling, severely limited the technology's application. Additionally, it has been theorized that fractionated resurfacing may be more effective because the pattern of thermal ablation allows for deeper dermal penetration. The greater depth of dermal penetration, which earlier ablative devices could not safely reach, is one explanation for the increased degree of tightness reported with FP. The fractionated CO_2 laser's quick recovery periods are a significant improvement over conventional CO_2 and Er:Yag laser resurfacing. It is believed that differences in wound healing processes are what cause the differences in healing times between standard ablative resurfacing and fractional resurfacing. Traditional ablative laser wound heals by migration of stem cells from the hair follicles [29].

Theoretically, re-epithelialization proceeds more rapidly with fractional ablative resurfacing due to the migration of adjacent cutaneous stem cells. To accurately define and explain the healing processes at work, more histologic and molecular study is needed. Unknown mechanisms may be at play when AFP improves and tightens skin. It has been demonstrated that after tissue is ablated, there is an initial bimodal contraction of the collagen and a second impact from the persistent collagen repair 3–6 months later [30]. Numerous fractionated CO_2 devices now available on the market allow for precise control of the amount and depth of cutaneous heating by having a range of fluences and pulse lengths. The intriguing potential of these devices to achieve changes in skin texture and laxity that are noticeably higher than the previous generation of non-ablative FP devices has been verified by clinical and histologic evidence. Recent research has assessed how using these gadgets affects photoaging. In 2009, studies conducted by the Food and Drug Administration utilizing an AFP device showed to have a high degree of safety and efficacy in photoaging, which resulted in the device's approval. The average improvement in rhytides, texture, and tightness indices on a quartile scale was 2.30, 2.42, and 1.65, respectively (laxity) [31].

A follow-up study on the long-term outcomes of ten patients who had previously undergone fractional CO_2 resurfacing was done by Ortiz et al. [32,119] Although clinical improvement was sustained over time, the results were not as spectacular as those seen at three-month follow-up visits. In comparison, participants retained 74% of their entire improvement during long-term visits. Although fractional CO_2 laser resurfacing has long-term effectiveness and durability of improvement of acne scarring and photodamage compared to baseline, more treatments may be necessary to maintain and/or enhance long-term outcomes. The authors hypothesized that the effects obtained at three months might be boosted by chronic inflammatory alterations, as demonstrated by heat shock protein 47 activation and continuous collagen remodeling seen in prior histologic investigations. Berlin et al. [33] conducted a research with ten individuals who underwent one treatment with an AFP device (UltraPulse Encore, Santa Clara, CA, USA). The authors investigated histologic and ultrastructural alterations in collagen deposition in addition to clinical changes in photoaging using light and electron microscopy. A study of skin textural changes and rhytid reduction performed by a blinded investigator revealed a mean improvement of 1.8 (on a five-point scale) after four weeks and 1.6 at 24 weeks after treatment. Increased fibrosis was seen in the papillary dermis in post-treatment biopsies. Additionally, electron imaging revealed that the average width of the collagen fibrils had decreased, correlating with an increase in the deposition of collagen type III, suggesting that new collagen had been deposited as a result of fractionated laser resurfacing. Karsai et al. [34] performed a rocessing controlled double-blind split-face trial to examine the effects of ablative fractional CO_2 and Er:Yag lasers on rhytides in the periorbital area. Previously, only ablative CO_2 laser resurfacing has been shown to enhance neck skin tightening. The examination included an evaluation of side effects and patient satisfaction in addition to a profilometric measurement of wrinkle depth and the Fitzpatrick wrinkle score [30]. While there was no obvious difference between the lasers, it is interesting to note that both modalities reduced periorbital wrinkle depth and Fitzpatrick score by around 20% and 10%, respectively. Pain and side effects were also considerably more noticeable after Er:Yag. Therapy in the early days, but following CO_2 laser therapy in the later course, there were more complaints. While many sessions are probably required for more widespread improvement, a single ablative fractional treatment session had a noticeable but limited effect on periorbital rhytides in the current investigation. Clinical-histologic correlation trials with FP have shown that a deeper injury increases texture improvement, but a higher density of more superficial injuries maximizes pigmentation improvement [35]. One of the factors contributing to the series' exceptional improvement in skin texture and laxity is the use of longer pulse durations (1000–1500 µs). Clinical and histologic correlation tests have shown via studies that a deeper depth of penetration of this AFP device at longer dwell times is associated with a greater improvement in tissue tightness. As a result, the pattern of deep ablation into the reticular dermis and subsequent activation of collagen contraction and synthesis are likely the causes of its greater effectiveness in tissue tightening and skin texture [36].

8.3.2.2 Solid State Lasers

While Nd: YAG (neodymium, yttrium, aluminum, and garnet) sources are the more well-known solid-state lasers, Nd: YLF (neodymium, yttrium, lithium, and fluoride) lasers are currently being investigated for alternative processing capabilities. Since the photons produced by both of these lasers have wavelengths in infrared region of the spectrum, the previous problems with thermal material contact would recur. Utilizing nonlinear crystals, solid-state lasers can be frequency converted to create photons at multiples of the fundamental frequency, allowing them to process materials at visible and ultraviolet wavelengths. The majority of processing work currently is done at fundamental frequencies, but as diode pumping of the rod technology rocessi, it is likely that more processing work will soon be done at frequency-converted wavelengths, particularly in the UV region, solid-state UV lasers will replace conventional UV sources for many applications [37]. Similar to CO_2 laser, today solid-state lasers are frequently used in focal-point configurations. The 1980s saw the introduction of excimer lasers, more recent photon sources, in industrial settings. Because they don't require frequency conversion, these laser sources directly produce photons with high average intensities in the UV band. Additionally, excimer lasers are used in an imaging configuration, allowing for multiple, simultaneous feature machining and straightforward beam splitting for multiple component processing due to the beam's incoherence and extreme divergence. Distinct spectral lines are released based on the gas mixture employed. Typically used for marking are longer-wavelength excimer lasers with wavelengths of 308 nm. This wavelength causes photochemical color changes in materials such as plastics and ceramics, as well as visible and permanent markings on some metals [37].

8.3.2.3 Fiber Lasers

Fiber lasers have big advantage over other lasers due to laser light being produced and delivered via a medium that is naturally flexible, making it simpler to deliver the laser light to the target and concentrating point. This can be crucial for metal and polymer laser cutting, welding, and folding. High output power in comparison to other laser types is another benefit. Since the active zones of fiber lasers can be very long, they can offer extremely high optical gain. The large surface area to volume ratio of the fiber, which enables effective cooling, enables them to maintain KW levels of continuous output power. The optical path's heat distortion is reduced or completely eliminated by the fiber's waveguide capabilities, leading to a typically high-quality, diffraction-limited optical beam. Because the fiber may be bent and coiled, with the exception of thicker rod-type systems, fiber lasers are smaller than solid-state or gas lasers of comparable power. They have a lower total cost of ownership, are dependable, display high thermal and vibrational stability, and have a longer lifetime [38–40]. High peak power and nanosecond pulses improve engraving and marking. The increased power and enhanced beam quality result in cleaner cut edges and higher cutting rates [27,28].

The processing of materials, communications, spectroscopy, medicine, and directed energy weapons are further used for fiber lasers [41–43]. The wavelength is the primary distinction that affects the kinds of materials each laser can treat. While CO_2 lasers have wavelengths in the 10,600 nm range, fiber lasers typically have wavelengths of 1,060 nm. In general, fiber lasers outperform CO_2 lasers in many ways.

8.3.2.4 Solid State Lasers in Medicines Today

Generally, 193 nm lasers are used for particular applications when other wavelengths are poorly absorbed by the material. These lasers are rarely frequently used because they have difficulties functioning at UV wavelengths less than 200 nm, including photon absorption in the air, a limited gas lifetime, poor laser output power, very high maintenance expenses, and the production of color centers in optics. Over 90% of all excimer laser applications employ 248 nm photons due to the aggressive material inter-action and extremely reliable laser operation at this particular wavelength. In most applications, feature resolutions of more than 10 μm are achievable, with a practical feature resolution of 1 μm. Equipment must be designed with additional care below 10 μm to reduce mechanical, thermal, vibrational, and optical aberrations. Excimer lasers are used in areas where CO_2 or solid-state sources are impractical due to the halogen gases they require, as well as their high cost and relative complexity of operation. Nonetheless, excimer lasers serve an essential processing niche and are usually the best technology available, which is capable of completing specific processes [44].

8.3.3 System Considerations

Once the proper laser has been selected, a whole processing system must include it. Finding the environment where the system will be installed is the first step. The system must be constructed to take use of the facilities that are available, especially when it comes to electrical and cleanroom needs as the majority of medical items need to be cleaned after processing. Additionally, it is required to choose whether the system will be operated online or as a standalone [45]. It takes in-depth expertise and comprehension of the operation of the line as a whole to integrate laser-based technologies into automated serial manufacturing lines, which frequently use conveyor-belt carriers. The production line should provide central management of all line stations and be constructed such that the product may move easily through each process. Three main functions are carried out by a laser beam delivery system. It propagates photons onto the workpiece first. In order to enhance the effectiveness of photon utilization, the system's optical components are employed to condition and shape the beam. This may entail beam homogeneity, beam mobility, or beam splitting (such as a galvanometer-driven head). Finally, the system must shield users from any potentially harmful surface reflections or light [46].

Most laser processing systems have a fixed beam, with the components moving rather than the beam. Aside from conveyor belt motion, additional phases can be employed to appropriately position the pieces during lasing. Dedicated tooling integrated into the motion control is used for rapid and precise part placement for actual part holding. When lasing a thin layer, vacuum chucks with solenoid control are often used, particularly in roll-to-roll applications. While certain production processes do not necessitate the use of cameras for visual inspection, component viewing of some kind is typically required or advised [34]. More advanced vision systems that take an image, digitize it, compare stored data to the target location, and align to specified fiducials can be chosen for highly automated and extremely exact alignment. Using a PC486 or an equivalent with the proper drivers, boards, and software, the motion system may be controlled by computer. CAD/CAM integration is a common feature. Modern production systems can be controlled using a keyboard, mouse, or touchscreen. The second feature is particularly helpful since it restricts access to the software for others by requiring just operator-level inputs on a touch screen, while senior engineers may fully configure the system using a keyboard and mouse. In order to assure operator safety, the whole system including regions for beam propagation, component positioning, and laser output; must be completely interconnected and enclosed. Excimer lasers need extra care since they employ high-pressure, halogen-containing gas combinations. It may also be useful to add further accessories such as gas processors, water chillers, and air-purification systems [47,124].

8.3.4 Processing Considerations

Disposable medical devices can also be made of glass, ceramics, and metals, although plastic of some kind is by far the most used material. Most disposable devices requiring LMM consist of catheters and injection-molded plastic components. Any component that needs to be processed must first undergo a thorough examination to determine how different laser sources interact with the material, while also taking into account the specifications for the size of the feature to be processed, the size of the processing area, the finish quality that is desired, and the anticipated production rate. Regarding etch rates for various materials. In several instances, laser processing can achieve the desired technical results, but due to cost constraints or inadequate throughput, or excessively large amortized piece costs, lasers cannot be used to achieve the desired results [47].

8.4 RECENT TRENDS IN BIOMEDICAL DEVICES

High-power lasers' ability to mold minute, fine components from surgical materials for implanted items such as stents, catheters, and needles has given medical applications a growing edge. Small features cannot be produced by conventional mechanical machining, which calls for several processes and tools. Chemical etching and electroforming procedures are restricted to a narrow set of metals with a simple (usually flat) structure, while another technique termed electrical discharge machining can only handle conductive materials. Plastic molding and metal casting are suitable for high-volume manufacture of items with comparatively lower accuracy, but they also need costly instruments. All of these methods' shortcomings are solved by LMM. It may be employed with practically any material, including metals and alloys, ceramics, polymers, multilayered materials, semiconductors, composites, and rubber. Only a few polymers, like Teflon, and transparent materials, like quartz, are challenging to treat with lasers without significantly increasing pulse energy. LMM is a one-stage, non-contact technique that enables feature accuracy and repeatability (hole diameter, slot width, etc.) within single microns. It is appropriate for high-volume manufacturing, low-volume manufacture of complicated components, and prototyping [47].

8.4.1 Bio-Implants

It is a huge issue for contemporary society to deal with diseases such as osteoarthritis and osteoporosis, which are prevalent in an older population. Bioimplants play a significant role in this. Bioimplants are leading the way in this regard. In the United States, the demand for primary complete hip and knee arthroplasties is anticipated to rise by 173% to 5,72,000 and 672% to 3,480,000 respectively, by the end of 2030, according to research by Kurtz et al. [48]. Given that this trend is likely to persist, it follows that advancements in bioimplant manufacturing will be required to meet the demand for production and the anticipated expansion [37].

The biomedical field's integration area has mostly been based on manufacturing throughout the past few decades. On the other hand, enterprises need to master modern manufacturing techniques appropriate for mass production. These criteria encourage a lot of industrial sector research and development in turn. Untreated bioimplants are vulnerable to wear and corrosion, two factors that are essential for having the best service life. It is crucial to offer sufficient biocompatibility after implantation. The host body normally responds to bioimplants in nanoseconds after initial contact, and the environment continues to alter beyond that. Poor biocompatibility will inevitably have substantial consequences such as immunologic rejection. To provide the bioimplants with unique surface qualities, a wide range of surface treatment methods are being explored [44]. At the present, biomaterials are frequently surface-modified before being used in practical applications. Surface coating on bioimplants frequently aims to improve wear and corrosion resistance. Strong osseointegration enables the ideal pace of degradation. A bioimplant's surface topography plays a key role in signaling cell function regulation and influences how the body will respond to the device. It has been found that several cell behaviors, including morphology, adhesion, orientation, migration, and differentiation, are all influenced by the textures or patterns on the surface. The alteration of surface topography with the goal of defining cells' reactivity has long been a study focus in the field of implantology since the biocompatibility of an implant is intimately linked to the response of cells in contact with the surface. The optimal surface roughness (R_a) for hard tissue implants, according to theoretical study, is in the range of 1 to 10 µm. Numerous in vivo and in vitro investigations have demonstrated that the best interlocking implant surface and mineralized bones are found in

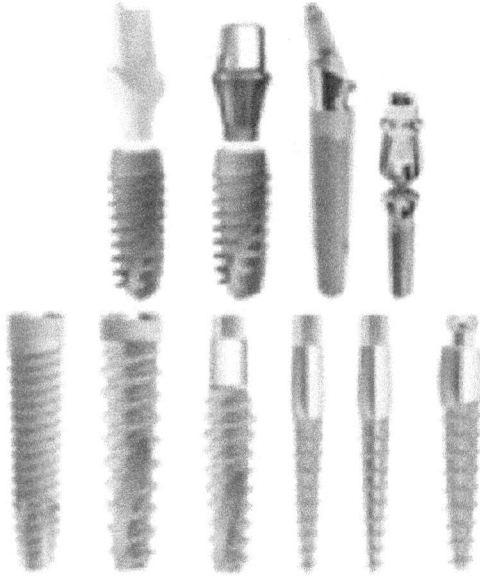

Figure 8.4 Dental implants made of Ti-6Al-4V and manufactured by LMM [49].

this roughness range. The microscale roughened surfaces in particular significantly induced osseointegration. In order to have a favorable impact on protein adsorption, cellular activity, and tissue responsiveness, appropriate surface modification technologies are being applied at the microscale. On the other hand, the majority of joint implants are experiencing tribology problems as a result of prolonged use. For instance, during walking movements, there would be a lot of rolling and sliding contacts at the prosthetic knee and hip joints. In the biomedical system, wear is often caused by friction between joint prosthesis, which increases energy losses [37]. Debris generated by wear would therefore cause negative immunological reactions as well as physical discomfort. Surface treatment is seen to be a potential way to lower the material friction coefficient and so extend the device lifetime rather than rectification by replacing the whole joint. In this context, surface texturing is preferred because it allows biomaterials to maintain their ideal bulk properties while also enhancing the tribological characteristics needed for various clinical applications [46]. Figure 8.4 shows the dental implants made up of Ti6Al4V using laser machining.

8.4.2 Surgical Tools

Minimally invasive surgery is one of the areas of the medical sector that is expanding the fastest. Particularly, intravenous treatments have increased as a result of three current medical objectives: lower costs, better patient outcomes, and faster recovery times. Tiny devices are essential to successful medical treatments, and UV LMM is the finest design and fabrication method for their unique forms and sizes. These devices must also be produced as cheaply as feasible because they are disposable rather than being implanted in the body.

No matter how big or small the project, an integrated laser workstation is required to manufacture the medical device design specifications in the best possible way. A workstation that uses excimer lasers substitutes a laser beam for a conventional cutting blade or drill bit. A line, rectangle, circle, hexagon, or other special forms can be created using a mask-projection or mask-scanning method. Any pattern can be moved into the beam path using an automatic mask changer, enabling on-the-fly machining.

Particularly, the market for minimally invasive medical devices has grown to include numerous new products in just the last few years. The fields of electrophysiology, stenting, and embolic protection are some of the ones in this device industry that are expanding the fastest.

Devices that use electrophysiology treat heart diseases, including arrhythmias. Micromachined devices are needed to implant cauterization and cryogenic catheters, which can kill tissue selectively. Some of the required laser procedures include excimer-processing operations like laser thinning or the removal of coating to provide electrical access to wires [50].

Each year, more and more stenting procedures are done. Stents were first placed into a vessel to stabilize it while it healed. However, the body occasionally reacted by developing restenosis or by enveloping the stent with scar tissue. The stent was coated with a biodegradable polymer saturated with an anti-restenosis medication to avoid this. Some businesses are creating a fully biodegradable stent that has been totally doped with an anti-restenosis chemical because the stent only needs to be in the body for three to six months.

Stents are utilized in conjunction with embolic protection devices. The successful creation of these devices is credited with the development of carotid artery stenting. Reopening a vessel can result in an embolic event because of the intervention and debris that results. Broken-off bits are prevented from entering the vessels and potentially triggering a stroke by an embolic filter. Excimer lasers can be used to laser drill filters.

Highly miniature 3D devices are essential for surgery to be as little intrusive as possible. The fabrication methods must advance as these gadgets' designs become ever more intricate and sophisticated. A potent technology for the microfabrication of these complex medical devices is excimer-LMM. The ongoing improvements in micromachining for the medical industry are necessary for life-saving and life-extending equipment [51].

8.4.3 Lab on Chip

Using technology that will enable improved dimensional precision and innovative designs in lab on a chip microfluidic biological and chemical species sensing devices is of great interest. Such lab on a chip microfluidic device can be made, in particular using laser processing technology. Different laser systems have also been used and optimized for processing of different glass and polymers over the past ten years. The use of femtosecond and picosecond laser systems has recently been applied to provide a cold type of ablation where the continuous internal microchannel ablation can also be performed due to a reduced associated HAZ, in addition to the conventional lasers used for processing glass and polymers. This chapter will provide a brief overview of the various laser fabrication techniques used to create microfluidic devices. Alternative common microfabrication techniques and materials will be briefly covered as well [52].

8.4.4 Bio Sensors

Sensor production is already a wide field in manufacturing and development. There are vast ranges of sensors from photodetectors to audio detectors and so much more. LMM is pivotal for producing highly sensitive detectors for just about any of these applications.

Femtosecond lasers are capable of incredible precision when it comes to machining materials. In many cases, stainless steel and optical fibers are the target materials. These materials are used in acoustic and optical sensors, respectively.

The carefully machined material surfaces using a laser allow for greater sensitivity and precision in these detectors, making micromachining crucial for development and production [53]. Femtosecond LMM of an compound parabolic concentrator fiber (CPC) tipped glucose sensors can be seen in Figure 8.5 [54].

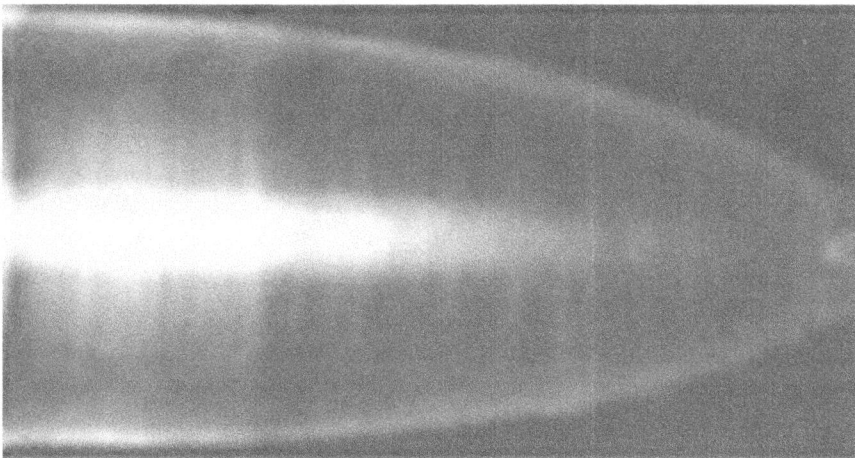

Figure 8.5 CPC tip glucose sensor (adapted from [54]).

8.4.5 Prosthetics

Ti-6Al-4V, an alpha-beta titanium alloy produced from LMM, is widely utilized in dental implants, hip and knee prostheses, and other medical equipment. Ti-6Al-4V implants are frequently treated with surface roughening and bioactive ceramic (such as hydroxyapatite) coatings to promote the formation of osteoblasts, the cells responsible for making bone [53,55]. Through the process of osseointegration, which involves the formation of bone tissue around an implant, these modified surfaces are employed to encourage implant fixation. The surface chemistry of implants may change as a result of conventional surface modification procedures, or debris may accumulate that could contribute to implant deterioration. On Ti-6Al-4V surfaces, Fasai et al. [56] created microscale grooves for cell growth and alignment using LMM.

A cochlear implant is a neuroprosthesis that is surgically installed and allows someone with moderate to extensive sensorineural hearing loss to perceive sound. Cochlear implants may enable better speech comprehension in both low- and high-volume contexts with the aid of treatment. Managing large ventral or incisional hernias can be challenging. Patients are typically sent to reconstructive surgeons following many unsuccessful hernia repair treatments. The treatment of complicated abdominal wall abnormalities has benefited substantially from the use of prosthetic and bioprosthetic materials which are manufactured by LMM [57–59].

For more than 50 years, prosthetic hip, knee, shoulder, and elbow joints have been anchored with great effectiveness using bone cements. Bone cement is used to anchor prosthetic joints, also known as artificial joints. The gap between the prosthesis and the bone is filled with bone cement, which also serves as an essential elastic zone. This is essential because the power exerted on the human hip by 10–12 times the body weight must be absorbed by the bone cement in order for the artificial implant to stay in place over the long term [60].

The purpose of fracture fixation therapy is to return the patient to normal function. Fixation of fractures can be achieved in a number of ways; one popular and effective method is fixation using screws and bone plates. Anatomical alignment and strong skeletal fixation are provided by bone plates. To maintain interfragmentary strain at a level consistent with fracture healing, such a stiff implant is used. The use of lasers in the precision machining of these screws and fixations results in a faultless fit [60]. Various types of implants in human anatomy are shown in Figure 8.6.

8.4.6 Angioplasty

Patients who have experienced a significant obstruction to blood artery in the circulatory system undergo angioplasty surgeries. The operation usually entails blowing up a balloon near the obstruction, which dislodges the

Cohlear Implants

Intacts

Cardiovascular Implants (Vascular Grafts)

Abdominal Wall Prosthesis

Prosthetic Arthroplasty

Intramedullary Nails

Knee joint Replacement, Tendon / Ligament, Cartilage Replacement

Dental Implants, Dental Post, Arch Wire & Brackets, Dental Bridges, Dental Restorative Material

Shoulder Prosthesis

Pacemaker

Lumbar Disc Replacement, Spine Cage, Plate, Rods and Screws

Total Hip Replacement, Acetabular

Bone Cement

Bone Fixation, Bone Plates & Screws

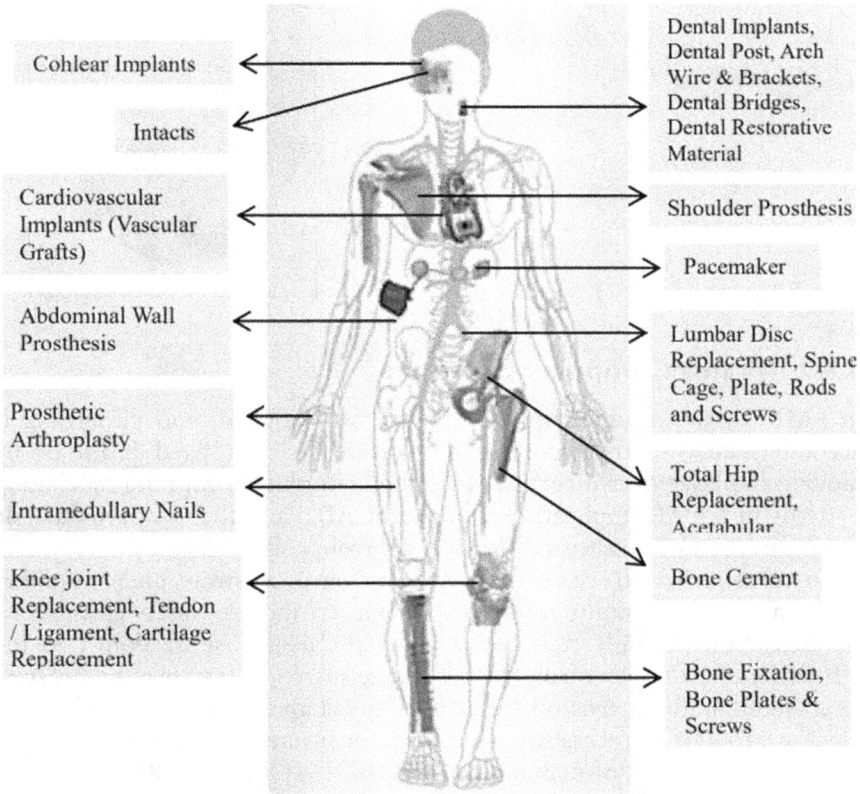

Figure 8.6 Different types of implants in human anatomy [61,62].

accumulated plaque and allows the channel to open. While this method is effective in the short term, 30 to 50% of angioplasty procedures require further care within six months. This is related to partial plaque clearance and restenosis, which is the creation of scar tissue as a result of vascular irritation. Because recurrent surgeries are costly, uncomfortable, and possibly fatal, there has been a significant drive in the healthcare sector in recent years to tackle restenosis. A number of the new technologies that medical device companies are creating heavily involve lasers.

The metal stent is one item that has attracted a lot of industry attention. These stents are now routinely inserted during angioplasty procedures since early outcomes appear encouraging. These stents are typically intricately carved to increase their flexibility while retaining the mechanical rigidity required to prevent the vessel walls from closing. Additionally, the usage of the devices helps to reduce the issue of arterial obstruction brought on by plaque entering the vessel following inflation [63]. The plaque is flattened due to inflated balloon as shown in Figure 8.7.

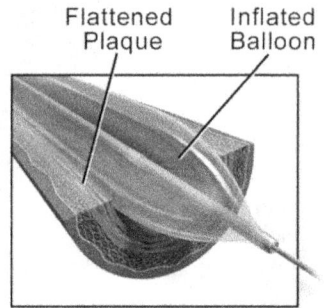

Figure 8.7 Angioplasty [57].

8.4.7 AI and Computer Software

In LMM, a tiny spot of the laser beam is collimated, and patterning is accomplished by either moving a substrate inside a fixed beam or by moving the laser over a surface. Simply create the appropriate machining patterns in a CAD application (like AutoCAD), and then import them as DXF file format into the LMM tool's control program [64].

To create relief structures or any holes on a substrate under ambient temperatures, commonly utilized procedures include laser cutting, or etching. With the only restrictions being the degree of freedom and the resolution of motion control, this technique's strength comes in its capacity to create desired patterns on surfaces of any shape. LMM is regarded as a quick prototyping process since it allows for instantaneous design changes without the creation of new molds or masks [65] (Figure 8.8).

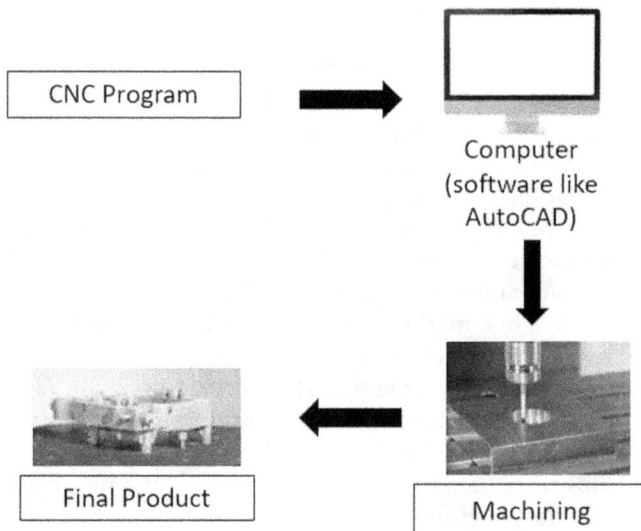

Figure 8.8 Use of software/computers in LMM.

8.5 CHALLENGES IN BIOMEDICAL INDUSTRY AND SOLUTION FOR FUTURE

One of the difficulties with laser materials processing is the removal of only the targeted material, typically by localized heating, but also decreasing the size of the HAZ (heat affected zone) to any residual material. To achieve this goal, it is important to precisely deliver laser irradiation with near-perfect beam quality to the target location. Higher quality outcomes can be attained with shorter wavelengths and smaller pulse widths.

For instance, most materials more readily absorb wavelengths in the Visible and UV spectral ranges, resulting in the shallow absorption depths and a greatly lower HAZ. For smaller, more precise machined features, UV lasers can also be focused into smaller places, and they have deeper depth of focus for better process yields. For immediate material vaporization, nonlinear absorption at the sample, very little heat transfer in material, and a negligible HAZ, ultrashort pulses in the fs (Femtoseconds) and ps (Picoseconds) regime produce strong maximum powers [23] (Figure 8.9).

Getting a high throughput of machining is a second issue for LMM. With some restrictions, higher average output power can result in higher ablation rates. When laser fluences (or energy densities) are outside of their ideal range, material removal efficiency is reduced. Insufficient fluence results in lower ablation rates, whereas excessive fluence reduces throughput and quality by partially depositing heat into the material. This is especially true for ultrashort pulses since maintaining the ideal operating regime for cold ablation depends on the delivery of the optimal fluence. Significant repetition rates with suitably high pulse energy are therefore required to produce the high output powers needed for improved throughput while keeping the best fluences for machining quality. Higher repetition rates also frequently lead to improvements in machining effectiveness and quality. The machine tool's other parts, however, may impose restrictions on the usable laser repetition rates. To benefit from the greater repetition rate, faster scanners and better motion control methods are required. In some circumstances, maintaining

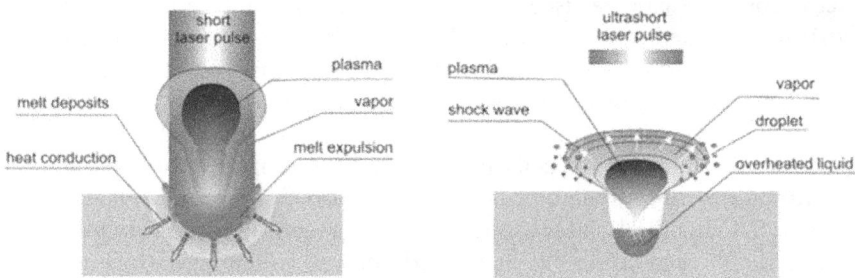

Figure 8.9 Impact of a laser pulse laser vs ultrashort pulse width on the machining quality [66].

appropriate fluence without scaling up repetition rate is possible by using spatial beam splitting and shaping techniques. However, these methods necessitate a more sophisticated system which is currently being created.

Another consideration for LMM in biomedical industry is cost. The most significant number may eventually be the cost difference between laser processing as compared to mechanical processing for each manufactured product. This metric takes into account the process yield, cost of operation, loss of productivity due to downtime, and upfront laser and system costs. Therefore, in addition to the initial cost, laser system dependability, service, and cost of service are crucial. Process yields losses immediately contribute to the system's operating costs as well. If the system has very high process stability over time and delivers reliable performance on a consistent basis, these expenses can be reduced [23].

Oncology is one area of medicine where laser therapy has the potential to significantly improve, both in terms of diagnosis and morbidity. Raman spectroscopy, a relatively new application of laser technology, has shown promising outcomes in the imaging of malignancy by discriminating between the light reflected back by normal and malignant tissue in the human body. It offers a non-intrusive imaging device [67–70].

With clinically based and laboratory investigations concentrating on malignancy-specific laser therapy, photodynamic therapy is an application that may one day give treatment to not only skin malignancies but also the gastro-esophageal, Genito-urinary, and respiratory tract neoplasia [71–74]. The use of photocoagulation, which may be a tissue-saving day treatment, to treat breast, liver, and benign tumors is also being investigated [71,75].

Theoretically sound and an area for further investigation, localized infections involving the destruction of bacterial and viral-infected cells can be treated using lasers [71,76].

In principle, if light and photosensitizing material can be given to the affected region, lasers may have a therapeutic impact against localized ill-nesses. This makes the use of lasers to treat diseases like MRSA-infected mouth ulcers or skin ulcers in the near future a very plausible possibility. Treatment for Helicobacter Pylori infection is another option [60]. Methylene blue can be used to photosensitize it, and endoscopic photo-dynamic treatment can be used to treat it. There is potential for sensitiza-tion or even the complete eradication of genital warts, despite the fact that treating viruses can be more difficult technically [71].

The development of the femtosecond laser has reduced the amount of light exposure that treated cells and tissues get and opened the possibility of general treatment with new lasers being developed to produce more precise dissection. The PIRL laser offers promising future clinical applications because of its capacity to cut tissue with almost no damage while avoiding coagulation and other types of tissue issues. This was determined when comparing the cutting effects of a scalpel, CO_2 laser, and the PIRL laser in cadaveric human vocal cords [77–82].

Compared to conventional procedures, laser-made microneedles and skin laser treatments enable the administration of intravenous and subcutaneous drugs without as much discomfort, improving drug delivery [83–86,120]. This is especially helpful for administering vaccinations since a cutaneous route frequently prompts a higher immune response than a muscle approach. Chen et al. [82] go through the benefit of employing lasers for vaccine administration, which can boost the immune response without having any noticeable local or systemic side effects. They contend that these lasers are an attractive subject for clinical trials in the near future due to their desired qualities, including ease, safety, and cost efficiency [83].

There are indications that low-level laser treatment can induce and expedite bone healing qualities in particular, leading to tissue regeneration (skin, bone, and cartilage) [87–92]. According to Renno et al.'s [93] vitro research on the effects of laser irradiation on osteoblasts and malignant osteosarcomas, various cell lines react differently to various wavelengths and energy levels. They showed that when exposed to an 830-nm laser, osteoblast proliferation significantly increased, however when exposed to a 780-nm laser, it decreased. According to Hawkins and Abrahamse [94,95], the right number of helium-neon laser exposures and energy density to injured fibroblasts can increase mitochondrial activity, which in turn can positively affect cellular migration and proliferation without putting additional stress on already compromised or damaged cells.

The option for robotic-assisted laser use is also an area gaining support [96,97]. Using lasers in laboratories is another topic. The capacity of laser imaging to highlight individual molecules and possibly target them for pharmacological therapy, as well as the possibility for lasers to hold and move minuscule units permitting individual cell separation, will only advance in the future. Another option is in vivo microscopy, which would again permit less tissue loss and improve the specificity of surgical treatments [98].

Intense pulse light sources (IPLs), which employ non-coherent light that spans the infrared to ultraviolet spectrum yet have laser-like characteristics, have recently undergone development. Currently, they are frequently used for hair removal, although they may have a wider range of applications [97,98].

In general, non-planar sample machining is becoming more and more important in many laser micro-machining applications, and both excimer laser mask projection system and direct write tools have already risen to the challenge. What to do with a sample of any shape, how to keep correct imaging or focus on the sample surface, and how to teach the system to micro-machine the required pattern accurately are all issues with this method, regardless of the technology that is taken into account. These technologies are in development stage for the micro-machining of 2.5D or 3D microstructures, as outlined in this study [99].

The use of lower excimer laser wavelengths of 157 nm, which can reach more materials with superior accuracy (for example, PTFE), and the use of ultra-short pulses, which can manufacture the largest variety of materials with the highest quality, are further areas of progress in laser micro-machining [99].

LMM is one of the technologies providing methodologies for advanced manufacturing and significant versatility in terms of materials to be micro-machined as well as the structures that can be tailored. Along with more traditional applications such as drilling, thin-film patterning, and laser marking, several new applications have arisen, primarily in the biomedical industry, that require more difficult operations like machining sophisticated surface profiles in the bulk material [100–106]. LMM has been developed to a stage now where more than one processing method is often suitable for any job. These developments have promoted laser-based technologies as they provide benefits in terms of technology, production, and the economy [107–115].

8.6 CONCLUSION AND FUTURE SCOPE

In biomedical industry the need of miniaturize devices is getting popularity. These miniature devices are difficult to manufacture and required advanced manufacturing techniques or micro-machining. Micromachining is state-of-the-art technology for precise production in the contemporary manu-facturing environment. A laser pulse is used in LMM to create intricate structures and tiny devices on any material that is now known to exist. In LMM, a laser beam is used as a cutting tool and the mechanism of machining is melting and vaporization. By use of targeted heating, melting, and subsequent vaporization, the material is removed in LMM. The process is governed by a number of criteria, all of which have a significant impact on the quality of the machining. Therefore, it is nec-essary to research the ideal criteria for getting the optimum machining quality. Currently there are very few research approaches, which have been published for developing biocompatible and antibacterial attributes together in the biomedical industry. In the future, we should expect to see innovative biomedical devices using artificial intelligence (AI) [122], machine learning (ML), and internet of things (IoT) that improve patient satisfaction.

Future advancements in laser technology could lead to the creation of new wavelength bands, increases in average output power, peak pulse en-ergy, and power levels, as well as reductions in cost, power consumption, and size efficiency. These guidelines will aid in the use of lasers in medicine across a larger range of specializations with easier, less expensive, and more compact laser applications, enabling laser used to be accepted more widely and its introduction to smaller hospital units.

Micromachining market growth in biomedical industry is expected to accelerate due to the increasing use of medical devices in emerging nations and the global increase in research into dissolvable and non-dissolvable medical micro-implants. By 2024, the medical sector is predicted to rule the micromachining market, expanding at a CAGR of 6.3% from 2019 to 2024. Businesses operating in these sectors are strategically expanding their product line.

Just by looking at these numbers, it can easily be concluded that there is so much potential for LMM in the Biomedical Industry. Thus, detailed investigations are required to identify the behavior of these innovative biomedical devices prior to clinical application, as well as a technique to increase biocompatibility (i.e., immediate biological responses). However, strong collaboration between surgeons, biologists, and engineers is required to meet the challenges of biomedical application.

REFERENCES

[1] Inchingolo, F., Hazballa, D., Inchingolo, A. D., Malcangi, G., Marinelli, G., Mancini, A., Maggiore, M. E., Bordea, I. R., Scarano, A., Farronato, M. J. M., 2022. Innovative concepts and recent breakthrough for engineered graft and constructs for bone regeneration: a literature systematic review. 15, 1120.

[2] Hsu, T.-R., 2008. *MEMS and microsystems: design, manufacture, and nanoscale engineering*. John Wiley & Sons.

[3] Bustillo, J. M., Howe, R. T., Muller, R.S.J.P.o.t.I., 1998. Surface micromachining for microelectromechanical systems. 86, 1552–1574.

[4] Maluf, N., Williams, K., 2004. *An introduction to microelectromechanical systems engineering*. Artech House.

[5] Tadigadapa, S. A., Najafi, N. J. J. M. S. E., 2003. Developments in microelectromechanical systems (MEMS): A manufacturing perspective. 125, 816–823.

[6] Tian, W.-C., Finehout, E., 2009. *Microfluidics for biological applications*. Springer Science & Business Media.

[7] Zhou, L., Fu, Q., Hu, D., Zhang, J., Wei, Y., Zhu, J., Song, J., Tong, M. J. C. S., 2021. A dense ZrB2-SiC-Si/SiC-Si coating to protect carbon/carbon composites against oxidation at 1773 K and 1973 K. 183, 109331.

[8] Masuzawa, T. J. C. A., 2000. State of the art of micromachining. 49, 473–488.

[9] Rizvi, N. H., Apte, P.J.J.o.M.P.T., 2002. Developments in laser micromachining techniques. 127, 206–210.

[10] Antimisiaris, S., Marazioti, A., Kannavou, M., Natsaridis, E., Gkartziou, F., Kogkos, G., Mourtas, S.J.A.d.d.r., 2021. Overcoming barriers by local drug delivery with liposomes. 174, 53–86.

[11] Razzacki, S. Z., Thwar, P. K., Yang, M., Ugaz, V. M., Burns, M.A.J.A.d.d.r., 2004. Integrated microsystems for controlled drug delivery. 56, 185–198.

[12] Sanjay, S. T., Zhou, W., Dou, M., Tavakoli, H., Ma, L., Xu, F., Li, X.J.A.d.d.r., 2018. Recent advances of controlled drug delivery using microfluidic platforms. 128, 3–28.

[13] Silvestri, S., Schena, E. J. M., 2012. Micromachined flow sensors in bio-medical applications. 3, 225–243.

[14] Caldorera-Moore, M., Peppas, N.A.J.A.d.d.r., 2009. Micro-and nano-technologies for intelligent and responsive biomaterial-based medical systems. 61, 1391–1401.

[15] Basak, S., Baumers, M., Holweg, M., Hague, R., Tuck, C. J. A. M., 2022. Reducing production losses in additive manufacturing using overall equipment effectiveness. 102904.

[16] Seliger, G., 2007. *Sustainability in manufacturing: recovery of resources in product and material cycles.* Springer.

[17] Palmberg, C., Dernis, H., Miguet, C., 2009. *Nanotechnology: an overview based on indicators and statistics.*

[18] Schaeffer, R., 2012. *Fundamentals of laser micromachining.* CRC press.

[19] Mishra, S., Yadava, V. J. O., engineering, l.i., 2015. Laser beam micro-machining (LBMM)–a review. 73, 89–122.

[20] Krüger, J., Kautek, W. J. P., Light, 2004. Ultrashort pulse laser interaction with dielectrics and polymers. 247–290.

[21] Khan Malek, C. G. J. A., chemistry, b., 2006. Laser processing for bio-microfluidics applications (part I). 385, 1351–1361.

[22] Zhang, Q., Shang, Q., Su, R., Do, T. T. H., Xiong, Q. J. N. L., 2021. Halide perovskite semiconductor lasers: materials, cavity design, and low threshold. 21, 1903–1914.

[23] Stratakis, E., Ranella, A., Fotakis, C., 2013. Laser-based biomimetic tissue engineering. 10.1007/978-3-642-41341-4_9

[24] Snoeys, R., Staelens, F., Dekeyser, W.J.C.a., 1986. Current trends in non-conventional material removal processes. 35, 467–480.

[25] Ceccarelli, F., Atzeni, S., Pentangelo, C., Pellegatta, F., Crespi, A., Osellame, R. J. L., Reviews, P., 2020. Low power reconfigurability and reduced crosstalk in integrated photonic circuits fabricated by femtosecond laser micromachining. 14, 2000024.

[26] Nijland, N., van Gemert-Pijnen, J., Boer, H., Steehouder, M., Seydel, E.J.J.o.m.I.r., 2008. Evaluation of internet-based technology for supporting self-care: problems encountered by patients and caregivers when using self-care applications. 10, e957.

[27] Zervas, M. N., Codemard, C. A., September 2014. High power fiber lasers: a review. *IEEE Journal of Selected Topics in Quantum Electronics.* 20(5), 219–241. Bibcode:2014IJSTQ..20..219Z. doi:10.1109/JSTQE.2014.23212 79. ISSN 1077-260X. S2CID 36779372.

[28] Phillips, K. C., Gandhi, H. H., Mazur, E., Sundaram, S. K., Dec 31, 2015). Ultrafast laser processing of materials: a review. *Advances in Optics and Photonics.* 7(4), 684–712. Bibcode:2015AdOP ... .7..684P. doi:10.1364/ AOP.7.000684. ISSN 1943-8206.

[29] Corkum, P.J.I.j.o.q.e., 1985. Amplification of picosecond 10 µm pulses in multiatmosphere CO 2 lasers. 21, 216–232.

[30] Garcin, C. L., Ansell, D. M., Headon, D. J., Paus, R., Hardman, M.J.J.S.c., 2016. Hair follicle bulge stem cells appear dispensable for the acute phase of wound re-epithelialization. 34, 1377–1385.

[31] Alexiades-Armenakas, M. R., Dover, J. S., Arndt, K.A.J.J.o.t.A.A.o.D., 2008. The spectrum of laser skin resurfacing: nonablative, fractional, and ablative laser resurfacing. 58, 719–737.

[32] Ortiz, A., Tremaine, A. & Zachary, C. (2010). Long-term efficacy of a fractional resurfacing device. *Lasers in surgery and medicine*. 42. 168–170. 10.1002/lsm.20885.

[33] Berlin, A. L., Hussain, M., Phelps, R., Goldberg, D. J., 2009. A prospective study of fractional scanned nonsequential carbon dioxide laser resurfacing: a clinical and histopathologic evaluation. *Dermatol Surg*. 35, 222–228.

[34] Karsai S., Czarnecka A., Ju¨nger M., Raulin C., 2010. Ablative fractional lasers (CO(2) and Er:YAG): a randomized controlled double-blind split-face trial of the treatment of peri-orbital rhytides. *Lasers Surg Med*. 42, 160–167.

[35] Grema, H., Greve, B., Raulin, C.J.L.i.S., Medicine, M.T.O.J.o.t.A.S.f.L., Surgery, 2003. Facial rhytides—subsurfacing or resurfacing? *A review*. 32, 405–412.

[36] Gallant, C. L., Olson, M. E., Hart, D. A. J. W. R., Regeneration, 2004. Molecular, histologic, and gross phenotype of skin wound healing in red Duroc pigs reveals an abnormal healing phenotype of hypercontracted, hyperpigmented scarring. 12, 305–319.

[37] Salvatore, S., Maggiore, U. L. R., Athanasiou, S., Origoni, M., Candiani, M., Calligaro, A., Zerbinati, N. J. M., 2015. Histological study on the effects of microablative fractional CO_2 laser on atrophic vaginal tissue: an ex vivo study. 22, 845–849.

[38] "Growing adoption of laser cutting machine market in the US through 2021, due to the need for superior-quality products: Technavio". Business Wire. Feb 2, 2017. Retrieved 2020-02-08.

[39] Shiner, B., Feb 1, 2016. Fiber lasers continue to gain market share in material processing applications. *SME.org*. Retrieved 2020-02-08.

[40] Shiner, B., Feb 1, 2006. High-power fiber lasers gain market share. *Industrial Laser Solutions for Manufacturing*. Retrieved 2020-02-08.

[41] Popov, S., 2009. 7: Fiber laser overview and medical applications. In Duarte, F. J. (ed.). *Tunable laser applications* (2nd ed.). New York: CRC.

[42] Frazier, A. B., Warrington, R. O., Friedrich, C., 1995. The miniaturization technologies: past, present and future. *IEEE Trans. on Industrial Electronics*, 42(5), 423.

[43] VIVA-Tech International Journal for Research and Innovation Volume 1, Issue 4 (2021) ISSN(Online): 2581–7280.

[44] Srinivasan, R. J. S., 1986. Ablation of polymers and biological tissue by ultraviolet lasers. 234, 559–565.

[45] Haveliwala, T. H., 2002. Topic-sensitive pagerank, Proceedings of the 11th international conference on World Wide Web, pp. 517–526.

[46] McClure, S. M., Laibson, D. I., Loewenstein, G., Cohen, J. D. J. S., 2004. Separate neural systems value immediate and delayed monetary rewards. 306, 503–507.

[47] Maier, M.W.J.S.E.T.J.o.t.I.C.o.S.E., 1998. Architecting principles for systems-of-systems. 1, 267–284.

[48] Kurtz, S., Ong, K., Lau, E., Mowat, F., Halpern, M., 2007. Projections of primary and revision hip and knee arthroplasty in the United States from 2005 to 2030. *The Journal of bone and joint surgery. American volume*, 89(4), 780–785. 10.2106/JBJS.F.00222

[49] Dziubińska, A., Majerski, K., Winiarski, G., 2017. Investigation of the effect of forging temperature on the microstructure of grade 5 titanium ELI. *Advances in Science and Technology Research Journal*. 11. 147–158. 10.12 913/22998624/76488.

[50] Kumar, A., Parihar, A., Basha, S. N., Panda, U., 2022 Computational approaches for novel therapeutic and diagnostic designing to mitigate SARS-CoV2 infection. pp. 451–488.

[51] Miller, P., Aggarwal, R., Doraiswamy, A., Lin, Y., Lee, Y.-S., Narayan, J., 2009. Laser micromachining for biomedical applications. *JOM*. 61, 35–40. 10.1007/s11837-009-0130-7

[52] Ben Azouz, A., Vazquez, M., Brabazon, D., 2014. Developments of laser fabrication methods for lab-on-a-chip microfluidic multisensing devices. *Comprehensive Materials Processing*. 13, 447–458. 10.1016/B978-0-08-096532-1.01317-0

[53] Maia, J. M., Amorim, V. A., Viveiros, D. et al., 2021. Femtosecond laser micromachining of an optofluidics-based monolithic whispering-gallery mode resonator coupled to a suspended waveguide. *Sci Rep* 11, 9128. 10.1038/s41598-021-88682-x

[54] Hassan, H. U.l., Lacraz, A., Kalli, K., Bang, O., 2017. Femtosecond laser micromachining of compound parabolic concentrator fiber tipped glucose sensors. *Journal of Biomedical Optics*. 22. 037003. 10.1117/1.JBO.22.3.037003

[55] Parmar, V., Kumar, A., Prakash, G. V., Datta, S., Kalyanasundaram, D., 2019 Investigation, modelling and validation of material separation mechanism during fiber laser machining of medical grade titanium alloy Ti6Al4V and stainless steel ss316l. *Mechanics of Materials*. 137, 103125.

[56] Fasasi, A. Y., Mwenifumbo, S., Rahbar, N., Chen, J., Li, M., Beye, A. C., Arnold, C. B., Soboyejo, W. O., 1 January 2009. Nano-second UV laser processed micro-grooves on Ti6Al4V for biomedical applications" *Materials Science and Engineering: C*. 29(1), 5–13.

[57] Blausen Medical Communications, Inc, 2013. Angioplasty—Balloon onflating in artery. *Wikimedia Commons*.

[58] Grevious, M. A. et al., 2006. The use of prosthetics in abdominal wall reconstruction. *Clinics in plastic surgery*. 33(2), 181–197, v. doi:10.1016/j.cps.2005.12.002

[59] Oshida, Y., Miyazaki, T., 2022. Chapter 7 Medical implants. *Biomaterials and engineering for Implantology: in Medicine and Dentistry*, De Gruyter: Berlin, Boston, pp. 229–300. 10.1515/9783110740134-008

[60] Parihar, A., Pandita, V., Kumar, A., Parihar, D. S., Puranik, N., Bajpai, T., Khan, R., 2021. 3D printing: Advancement in biogenerative engineering to combat shortage of organs and Bioapplicable materials. *Regen. Eng. Transl. Med.*, 1–27.

[61] Patel, N. R., Gohil, P. P., 2012. A review on biomaterials: scope, applications & human anatomy significance.

[62] Parmar, V., Kumar, A., Mani Sankar, M., Datta, S., Vijaya Prakash, G., Mohanty, S., Kalyanasundaram, D., 2018. Oxidation facilitated antimicrobial ability of laser micro-textured titanium alloy against grampositive Staphylococcus aureus for biomedical applications. *Journal of Laser Applications*. 30(3), 032001.

[63] Ramanathan, D., 2012, December 1. Advantages of laser micromachining for balloons and catheters. *Medical Design Briefs.*

[64] Schaeffer, R. D., 2012. *Fundamentals of laser micromachining.* CRC Press: Boca Raton, FL, USA.

[65] Yeh, J. T. C., 1986. Laser ablation of polymers. *J. Vac. Sci. Technol. A.* 4, 653.

[66] Kleine, K., LaHa, M., 2019, May 21. Femtosecond lasers for unmatched micromachining. *Equipment news.*

[67] Conzemius, M., Swainson, S., 1999. Fracture fixation with screws and bone plates. *The Veterinary clinics of North America. Small animal practice.* 9(5), 1117–1133, vi. doi:10.1016/s0195-5616(99)50105-8

[68] Canetta E., Riches A., Borger E., Herrington S., et al., Discrimination of bladder cancer cells from normal urothelial cells with high specificity and sensitivity: Combined application of atomic force microscopy and modulated Raman spectroscopy Acta Biomater 2014; pii: S1742-7061(13): 00662-4. Jan 7.

[69] Kast, R. E., Tucker, S. C., Killian, K., Trexler, M., Honn, K. V., Auner, G. W., 2014. Emerging technology: applications of Raman spectroscopy for prostate cancer. *Cancer Metastasis Rev.* 33(2-3), 673–693.

[70] Kallaway, C., Almond, L. M., Barr, H., et al., 2013. Advances in the clinical application of Raman spectroscopy for cancer diagnostics. *Photodiagn Photodyn Ther.* 10(3), 207–219.

[71] Brown, S. G., 1998. Science, medicine, and the future: new techniques in laser therapy. *BMJ.* 316, 754–757.

[72] Deprez, P. H., 2014. Future directions in EUS-guided tissue acquisition. *Gastrointest Endosc Clin N Am.* 24(1), 143–149.

[73] Crous, A. M., Abrahamse, H., 2013. Lung cancer stem cells and low-intensity laser irradiation: a potential future therapy?. *Stem Cell Res Ther.* 4(5), 129.

[74] Valcavi, R., Piana, S., Bortolan, G. S., Lai, R., Barbieri, V., Negro, R., 2013. Ultrasound-guided percutaneous laser ablation of papillary thyroid micro-carcinoma: a feasibility study on three cases with pathological and immu-nohistochemical evaluation. *Thyroid.* 23(12), 1578–1582.

[75] Ashiq, M. G., Saeed, M. A., Tahir, B. A., Ibrahim, N., Nadeem, M., 2013. Breast cancer therapy by laser-induced Coulomb explosion of gold nano-particles. *Chin J Cancer Res.* 25(6), 756–761.

[76] Raab, O., 1900. Ueber die wirkung fluorescierenden stoffe auf infusiorien. *Z Biol.* 39, 524–546.

[77] Tracey, S., Gracco, A., 2012. Lasers in Orthodontics. In: Graber, L. W., Vanarsdall, R. L., Vig, K. W. L., Eds. *Orthodontics current principles and techniques.* 5th ed. Elsevier Mosby: St. Louis, Mo, pp. 1051–1073.

[78] Abrahamse, H., 2012. Regenerative medicine, stem cells, and low-level laser therapy: future directives. *Photomed Laser Surg.* 30(12), 681–682.

[79] Roberts, T. V., Lawless, M., Chan, C. C., et al., 2013. Femtosecond laser cataract surgery: technology and clinical practice. *Clin Exp Ophthalmol.* 41(2), 180–186.

[80] Sutton, G., Bali, S. J., Hodge, C., 2013. Femtosecond cataract surgery: transitioning to laser cataract. *Curr Opin Ophthalmol.* 24(1), 3–8.

[81] Kuetemeyer, K., Rezgui, R., Lubatschowski, H., Heisterkamp, A., 2010. Influence of laser parameters and staining on femtosecond laser-based intracellular nanosurgery. *Biomed Opt Express.* 1(2), 587–597.

[82] Hess, M., Hildebrandt, M. D., Müller, F., et al. Picosecond infrared laser (PIRL): an ideal phonomicrosurgical laser? *Eur Arch Otorhinolaryngol* 2013; 270(11): 2927–2937.

[83] Chen, X., Wang, J., Shah, D., Wu, M. X., 2013. An update on the use of laser technology in skin vaccination. *Expert Rev Vaccines.* 12(11), 1313–1323.

[84] Scheiblhofer, S., Thalhamer, J., Weiss, R., 2013. Laser microporation of the skin: prospects for painless application of protective and therapeutic vaccines. *Expert Opin Drug Deliv.* 10(6), 761–773.

[85] Elsabahy, M., Foldvari, M., 2013. Needle-free gene delivery through the skin: an overview of recent strategies. *Curr Pharm Des.* 19(41), 7301–7315.

[86] Yavlovich, A., Smith, B., Gupta, K., Blumenthal, R., Puri, A., 2010. Light-sensitive lipid-based nanoparticles for drug delivery: design principles and future considerations for biological applications. *Mol Membr Biol.* 27(7), 364–381.

[87] Meier, J. C., Bleier, B. S., 2013. Novel techniques and the future of skull base reconstruction. *Adv Otorhinolaryngol.* 74, 174–183.

[88] Bleier, B. S., Cohen, N. A., Chiu, A. G., Omalley, B. W. Jr, Doghramji, L., Palmer, J. N., 2010. Endonasal laser tissue welding: first human experience. *Am J Rhinol Allergy,* 24(3), 244–246.

[89] Matteini, P., Ratto, F., Rossi, F., Pini, R., 2012. Emerging concepts of laser-activated nanoparticles for tissue bonding. *J Biomed Opt.* 17(1), 010701.

[90] Capon, A., Mordon, S., 2003. Can thermal lasers promote skin wound healing? *Am J Clin Dermatol.* 4(1), 1–12.

[91] Sobol, E., Shekhter, A., Guller, A., Baum, O., Baskov, A., 2011. Laser-induced regeneration of cartilage. *J Biomed Opt.* 16(8), 080902.

[92] Esposito, G., Rossi, F., Matteini, P. et al., 2011. Present status and new perspectives in laser welding of vascular tissues. *J Biol Regul Homeost Agents.* 25(2), 145–152.

[93] Renno, A. C., McDonnell, P. A., Parizotto, N. A., Laakso, E. L., 2007. The effects of laser irradiation on osteoblast and osteosarcoma cell proliferation and differentiation in vitro. *Photomed Laser Surg.* 25(4), 275–280.

[94] Hawkins, D., Abrahamse, H., 2006. Effect of multiple exposures of low-level laser therapy on the cellular responses of wounded human skin fibroblasts. *Photomed Laser Surg.* 24(6), 705–714.

[95] Kumar, S., Verma, R. K., Kumar, A., Patel, V. K., 2022. Importance of chemically treated natural fibers in the fabrication of natural fiber reinforced polymer composites. In *Trends in fabrication of polymers and polymer composites*, AIP Publishing Books, pp. 10-1–10-22.

[96] Gonzalez-Martinez, J., Vadera, S., Mullin, J., et al., 2014. Robot-assisted stereotactic laser ablation in medically intractable epilepsy: operative technique. *Neurosurgery.* 10(Suppl. 2), 167–172.

[97] Jolesz, F. A., 2011. Intraoperative imaging in neurosurgery: where will the future take us? *Acta Neurochir Suppl (Wien).* 109, 21–25.

[98] Mooney, M. A., Zehri, A. H., Georges, J. F., Nakaji, P., 2014. Laser scanning confocal endomicroscopy in the neurosurgical operating room: a review and discussion of future applications. *Neurosurg Focus.* 36(2), E9.

[99] Rizvi, N. H., Apte, P., 2002. Developments in laser micro-machining techniques. *Journal of Materials Processing Technology,* 127(2), 206–210, ISSN 0924-0136.

[100] Gangwar, A. K. S., Rao, P. S., Kumar, A., 2021. Bio-mechanical design and analysis of femur bone. *Materials Today: Proceedings*. 44(Part 1), 2179–2187, ISSN 2214-7853. 10.1016/j.matpr.2020.12.282

[101] Gangwar, A. K. S., Rao, P. S., Kumar, A., Patil, P. P., 2019. Design and Analysis of Femur Bone: BioMechanical Aspects. *Journal of Critical Reviews*, 6(4), 133–139, ISSN-2394-5125.

[102] Kumar, A., Mamgain, D. P., Jaiswal, H., Patil, P., 2015. Modal analysis of hand arm vibration (humerus bone) for biodynamic response using varying boundary conditions based on FEA. *Advances in Intelligent Systems and Computing*. 308, 169–176. DOI: 10.1007/978-81-322-2012-1_18

[103] Kumar, A., Behmad, S. I., Patil, P., 2014. Vibration characterization and static analysis of cortical bone fracture based on finite element analysis. *Engineering and Automation Problems*, No 3-2014, pp. 115–119. UDC- 621.

[104] Kumar, A., Jaiswal, H., Garg, T., Patil, P., 2014. Free vibration modes analysis of femur bone fracture using varying boundary conditions based on FEA. *Procedia Materials Science*. 6, 1593–1599. DOI: 10.1016/j.mspro.2014.07.142

[105] Kumar, A., Gori, Y., Rana, S., Sharma, N. K., Yadav, B., 2022. FEA of Humerus Bone Fracture and Healing. *Advanced materials for biomechanical applications* (1st ed.) CRC Press, 10.1201/9781003286806

[106] Kumar, A., Datta, A., Kumar, A., 2022. Recent Advancements and Future Trends in Next-Generation Materials for Biomedical Applications. In Kumar, A., Gori, Y., Kumar, A., Meena, C. S., Dutt, N., Eds., *Advanced materials for biomedical applications*. CRC Press: Boca Raton, FL, USA, Chapter 1, pp. 1–19.

[107] Prasad, A., Chakraborty, G., Kumar, A., 2022. Bio-based Environmentally Benign Polymeric Resorbable Materials for Orthopedic Fixation Applications. In Kumar, A., Gori, Y., Kumar, A., Meena, C. S., Dutt, N., Eds., *Advanced materials for biomedical applications*. CRC Press: Boca Raton, FL, USA, Chapter 15, pp. 251–266.

[108] Datta, A., Kumar, A., Kumar, A., Kumar, A., Singh, V. P., 2022. Advanced Materials in Biological Implants and Surgical Tools. In Kumar, A., Gori, Y., Kumar, A., Meena, C. S., Dutt, N., Eds., *Advanced materials for biomedical applications*. CRC Press: Boca Raton, FL, USA, Chapter 2, pp. 21–43.

[109] Kumar, A., Gangwar, A. K. S., Kumar, A., Meena, C. S., Singh, V. P., Dutt, N., Prasad, A., Gori, Y., 2022. Biomedical Study of Femur Bone Fracture and healing. In Kumar, A., Gori, Y., Kumar, A., Meena, C. S., Dutt, N., Eds., *Advanced materials for biomedical applications*. CRC Press: Boca Raton, FL, USA, Chapter 14, pp. 235–250.

[110] Patil, P. P., Gori, Y., Kumar, A., Tyagi, M., 2021. Experimental analysis of tribological properties of polyisobutylene thickened oil in lubricated contacts. *Tribology International*. 159, 106983.

[111] Kumar, A., Rana, S., Gori, Y., Sharma, N. K., 2021. Thermal Contact Conductance prediction Using FEM Based computational Techniques. In *Advanced computational methods in mechanical and materials engineering*. CRC Press: Boca Raton, FL, USA, pp. 183–220. ISBN 9781032052915.

[112] Kumar, A., Patil, P. P., 2016. FEA simulation and RSM based parametric optimization of vibrating transmission gearbox housing. *Perspect. Sci.* 8, 388–391.

[113] Patil, P. P., Sharma, S. C., Jaiswal, H., Kumar, A., 2014. Modeling influence of tube material on vibration based EMMFS using ANFIS. *Procedia Mater. Sci.* 6, 1097–1103.

[114] Bhoi, S., Kumar, A., Prasad, A., Meena, C. S., Sarkar, R. B., Mahto, B.,Ghosh, A., 2022. Performance evaluation of different coating materials in delamination for micro-milling applications on high-speed steel substrate. *Micromachines*, 13, 1277. 10.3390/mi13081277

[115] Bhoi, S., Prasad, A., Kumar, A., Sarkar, R. B., Mahto, B., Meena, C. S., Pandey, C., 2022. Experimental study to evaluate the wear performance of UHMWPE and XLPE material for orthopedics application. *Bioengineering.* 9, 676. 10.3390/bioengineering9110676

[116] Parihar, Arpana, Kumar, Avinash, Panda, Udwesh, Khan, Rukhsar, Parihar, Dipesh Singh, & Khan, Raju (2023). Cryopreservation: A Comprehensive Overview, Challenges, and Future Perspectives. *Advanced Biology*, 2200285. 10.1002/adbi.202200285

[117] Kumar, Avinash, Sharma, Anuj Kumar, & Katiyar, Jitendra Kumar (2023). State-of-the-Art in Sustainable Machining of Different Materials Using Nano Minimum Quality Lubrication (NMQL). *Lubricants*, 11, 64. 10.3390/lubricants11020064.

[118] Subramanian, Yathavan, Gajendiran, J., Veena, R., Azad, Abul Kalam, Sabarish, V. C. Bharath, Muhammed Ali, S. A., Kumar, Avinash, & Gubendiran, Ramesh Kumar (2023). Structural, Photoabsorption and Photocatalytic Characteristics of BiFeO3-WO3 Nanocomposites: An Attempt to Validate the Experimental Data Through SVM-Based Artificial Intelligence (AI). *Journal of Electronic Materials*, 52, 2421–2431. 10.1007/s11664-022-10188-7.

[119] Kumar, Avinash, Parihar, Arpana, Panda, Udwesh, & Parihar, Dipesh Singh (2022). Microfluidics-Based Point-of-Care Testing (POCT) Devices in Dealing with Waves of COVID-19 Pandemic: The Emerging Solution. *ACS Applied Bio Materials*, 5, 2046–2068. 10.1021/acsabm.1c01320.

[120] Kumar, Avinash, Shrama, Anuj Kumar, Gupta, TVK, & Katiyar, Jitendra Kumar (2022). Influence of hexagonal boron nitride additive nanocutting fluid on the machining of AA6061-T6 alloy using minimum quality lubrication. *Proceedings of the Institution of Mechanical Engineers, Part E: Journal of Process Mechanical Engineering*, 095440892211109. 10.1177/09544089221110980

[121] Chakraborty, Gourhari, Prasad, Arbind, & Kumar, Ashwani (2022). *Processing of Biodegradable Composites" Biodegradable Composites for Packaging Applications*. CRC Press, Taylor & Francis, ISBN: 978103227908. DOI: 10.1201/9781003227908-3.

[122] Subramanian, Yathavan, Veena, R., Muhammed Ali, S.A., Kumar, Avinash, Gubediran, Ramesh Kumar, Dhanasekaran, Anitha, Gurusamy, Dhanasekaran, & Muniandi, Kalpana (2021). Artificial intelligence technique based performance estimation of solid oxide fuel cells. *Materials Today: Proceedings*. 10.1016/j.matpr.2021.06.412

[123] Singh, Manjinder, Kumar, Avinash, & Khan, Abdul Rahman (2020). Capillary as a liquid diode. *Physical Review Fluids*, 5. 10.1103/physrevfluids.5.102101.

[124] Kumar, Avinash, Datta, Subhra, & Kalyanasundaram, Dinesh (2017). Liquid Slippage in Confined Flows: Effect of Periodic Micropatterns of Arbitrary Pitch and Amplitude. *Journal of Heat Transfer*, 140. 10.1115/1.4037363.

Effect of Laser Surface Melting on Atmospheric Plasma Sprayed High-Entropy Alloy Coatings

Himanshu Kumar, S.G.K. Manikandan, M. Kamaraj, and S. Shiva

CONTENTS

9.1   Introduction 207
9.2   Laser-Based Surface Modification Techniques 209
     9.2.1   Laser Surface Texturing 210
     9.2.2   Laser Cladding 211
     9.2.3   Laser Shock Peening..212
     9.2.4   Laser Surface Melting..213
     9.2.5   Laser Hardening...213
     9.2.6   Laser Surface Alloying...215
     9.2.7   Laser Glazing 216
9.3   Laser–Material Interaction 216
9.4   Area of Applications 217
     9.4.1   Automobile Applications..218
     9.4.2   Aerospace Applications 219
     9.4.3   Biomedical Applications 220
     9.4.4   Nuclear Applications..220
9.5   High-Entropy Alloy..220
9.6   Experimental Procedure 223
9.7   Results and Discussion..226
9.8   Conclusions...227
9.9   Future Scope 228
Acknowledgments..229
References...229

9.1 INTRODUCTION

A novel type of material called high-entropy alloys (HEAs) is comprised of elements that are almost or precisely equiatomic in ratio, with 5 at. % to 35 at. % in terms of the chemical constituents [1,2]. The HEAs offer enhanced microhardness, wear, and corrosion resistance abilities compared to conventional alloys [3–5]. To some extent, HEAs are emerging as the next

DOI: 10.1201/9781003402398-9

generation of coating options. Numerous HEA coatings/films have been successfully manufactured by various methods such as vacuum arc melting [6], cold spraying [7], high-velocity oxy-fuel spraying [8], wire arc additive manufacturing [9], and magnetron sputtering [10]. However, wear applications need surface quality enhancements, such as decreased roughness, enhanced hardness, and the elimination of surface flaws, to prolong the component's useful life. In order to enhance the surface properties, various laser-based surface modification techniques such as laser cladding (LC) [11], laser hardening (LH) [12], laser shock peening (LSP) [13], laser surface melting (LSM) or laser re-melting [14], laser surface alloying (LSA) [15], laser glazing (LG) [16], and laser surface texturing (LST) [17] were developed in the past decade.

In general, LSM is a promising surface modification technique to improve the hardness, wear, corrosion, and fatigue properties. It is widely used for steels, aluminum alloys, zirconia-based alloy, nickel base alloy, and copper–chromium alloy [18–22]. Liu et al. used LSM on DH36 steel; results indicate the improvement in the yield strength and tensile strength; however, the hardness is decreased. The strengthening mechanism involved in the process is second phase strengthening and solid–solution strengthening at 1 kW and 2.5 kW laser power, respectively [18]. Rodríguez et al. used the LSM process to melt the pre-mixed CNT-Al powder on Al 6061substrate; results revealed a higher hardness value and greater homogeneity [19]. Furthermore, Jiao et al. used the nano pulsed laser for surface melting of Zr-based bulk metallic glasses in order to enhance the shear bonding. The results revealed that LSM-induced surface hardening was associated with the crystalline precipitates and the existence of compressive residual.

On the other hand, with a softened surface post-LSM, tensile residual stress and a decreased proportion of crystalline precipitates were observed. However, variations in shear-banding processes were seen close to the surface of the laser-irradiated zones [20]. The behavior of epitaxial growth and the mechanism of stray grains production during laser surface re-melting of Ni-based alloy were investigated by Du et al. [21]. Findings revealed that element segregation triggers the appearance of stray grains near eutectic and carbides phases on the fusion line, about 50 J/mm of heat is what's needed to keep the stray grain under control. These results may be helpful for the reclamation of directionally solidified superalloys that cannot be welded. Zhang et al. used Nd: YAG laser for LSM to process the Cu-50Cr alloy. The results show the significant grain refinement from ~100 µm to ~1 µm, the segregation of the liquid phase was inhibited using rapid cooling, and the electrical and mechanical properties of the alloys were enhanced [22].

Numerous industrial sector relies increasingly on LSM methods for engineering applications. Several lasers, including CO_2, Nd: YAG, fiber, and excimer lasers, are used in the LSM process. Factors including laser system availability, maintenance costs, and efficiency should all be considered when choosing a laser source. Rakesh et al. used the LSM process on Mg-Zn-Dy alloy to improve the implant application's corrosion resistance and wettability [23]. The investigation shows that the grain is refined and the microhardness is much enhanced. A decrease in grain size and an increase in roughness are responsible for the enhanced wettability. LSM is also used for polishing the surface using an ultra pulse laser by Astrid et al. Findings show a decrease in the surface roughness value from 0.6 μm to 0.3 μm [24]. LSM was performed on the Cr–Fe–Co–Ni–Al duel phase alloy using 180 W Nd: YAG laser. The results revealed that $CrFeCoNiAl_{0.6}$ and $CrFeCoNiAl_{0.7}$ alloys exhibit superior corrosion properties and enhanced hardness [25].

Furthermore, laser re-melted Zr-Nb-V-Ti-Al alloy exhibits the refined grain with homogeneity in composition [26]. $CoCrFeNiAl_1$ alloy synthesized using the re-melting exhibits dense, uniform, and superior high-temperature oxidation resistance [27]. Similarly, laser re-melted $FeCoCrNiAl_{0.5}Si_2$ alloy revealed that phase transformation from FCC to BCC leads to enhanced microhardness and wear resistance [28]. Correspondingly, laser re-melted $CoCrFeNiAl_{0.5}Ti_{0.5}$ alloy exhibits superior microhardness and wear resistance post-treatment [29]. Cai et al. used laser re-melting to treat FeMnCrNiCo + 20%TiC alloy; results indicate composition uniformity and enhanced wear resistance [30]. However, Chai et al. treated CrCoNi alloy with LSM and exhibited a decreased hardness value attributed to grain coarsening [31]. NiCoCrA1Y thermal barrier coating exhibits superior surface qualities post pulse laser treatment [32]. Novak et al. produced $Al_{0.5}CoCrCuFeNi$ alloy using laser melting showing a micro-hardness of 151 ± 11 HV [33]. Compositional segregation could be hampered, solubility in solid–solution phases could be enhanced, and the strengthening improved by grain refinement if laser-induced fast solidification was used [34]. Similarly, laser re-melted FeCoCrNiAl alloy exhibits superior wear resistance [35].

As mentioned above, the literature shows that HEA exhibits superior microhardness, wear, and corrosion resistance. The current chapter discusses the various types of laser surface processing and their applications. Furthermore, the effects of LSM on the microstructure, microhardness, and density of CoCrFeNiW HEAs are discussed in detail.

9.2 LASER-BASED SURFACE MODIFICATION TECHNIQUES

Laser material processing modifies the surface characteristics of metals, ceramics, glasses, and polymers. Many of these processes include heating, although they are not exclusively thermal. Laser surface modification technology includes laser texturing, LC, LSP, laser re-melting/laser surface processing, LH, laser alloying, and LG. Various types of laser-based surface modification techniques are shown in Figure 9.1. Laser beams interact directly with the workpiece and sometimes material applied to the surface.

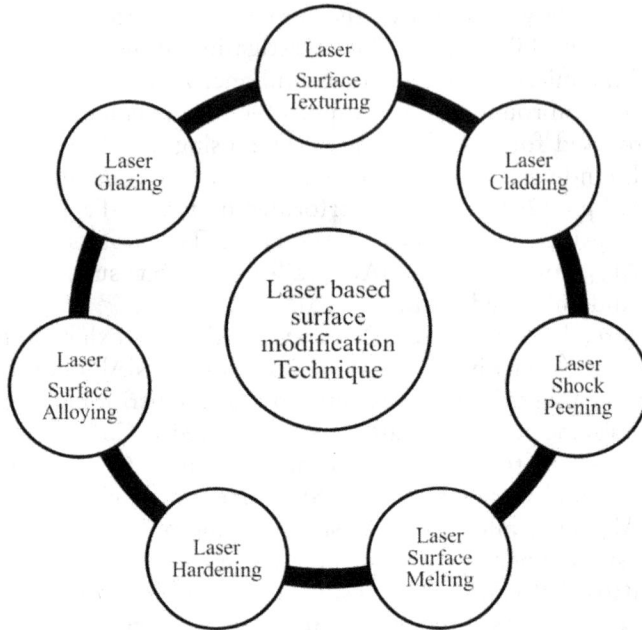

Figure 9.1 Laser-based surface modification techniques.

There are several types of lasers, each with its own pros and cons for processing the materials. The wavelength of the laser beam has a significant effect on the heat transfer to the materials.

In most cases, when a laser is used to process material, that laser heat energy may be broken down into two categories: the portion of the laser's rays that are absorbed by the work material and the portion that is reflected back into the surrounding environment. Laser heat energy transferred to the work material primarily depends on work material heat absorbability and irradiation wavelength. In this section, various types of laser-based surface modification techniques are discussed in detail.

9.2.1 Laser Surface Texturing

LST, also known as laser ablating, is a method of texturing the surface by selectively removing the material. It's capable of producing precise micro/nanopatterns with higher repeatability. LST gained attention over the traditional process due to its higher precision, repeatability, reliability, cost-effective, and noncontact process. It is extensively used to improve the surface properties such as wettability, adhesion, friction, and electrical and thermal conductivity. The LST process comprises a laser source, a focusing lens, and an integrated computer system for texture design. A schematic diagram of the LST process is shown in Figure 9.2. Various patterns such as

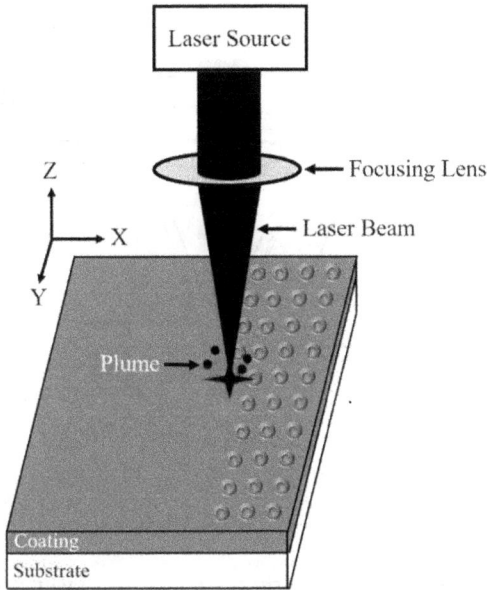

Figure 9.2 Schematic layout of LST process.

dimples, grooves, and complex shape textures can be created by adjusting the scanning pattern, laser power, and scanning rate. LST is widely used in mechanical sealing, microelectromechanical systems devices, conducting materials, solar cells, and dental and bone implants. Recently, LST has been widely used to produce a hydrophobic coating.

9.2.2 Laser Cladding

LC is used to apply a thick coating over the surface to protect from harsh environmental conditions and sometimes in the reclamation of the components. LC utilized a high-power laser source to clad the material over the substrate. Powder and wire are widely used as cladding materials. Thin-wall samples undergo less distortion during the heating and solidification stages of the LC process, and their reduced heat affected zone, rapid heating, and solidification contribute to microstructural changes. Figure 9.3 represents a schematic diagram of the LC process. It consists of a laser source, optics, powder feeder, and shielding gas nozzle. Generally, argon (Ar) gas is used to feed the powder as well as shielding gas. LC has three main advantages: superior surface homogeneity, minimal porosity, and low dilution (about 10%). It is widely used in wear, corrosion, and aerospace applications. Compared to conventional welding methods, LC has a few drawbacks, including a more significant initial investment, a cooling system, optics, and an integrated computer system. One may achieve the desired property by

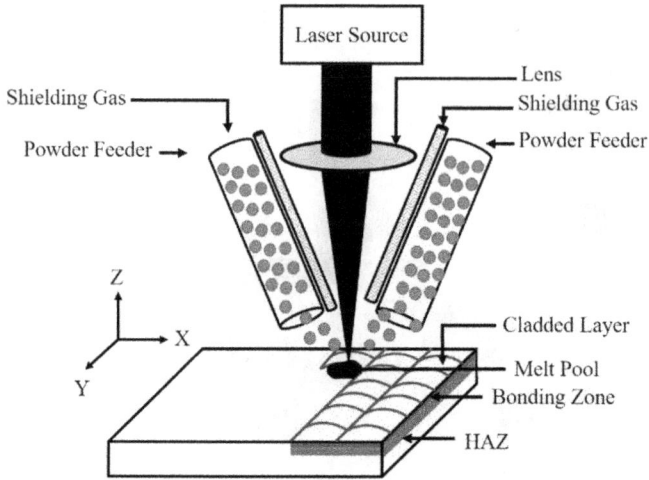

Figure 9.3 A schematic representation of the LC process.

adjusting the laser process parameters, such as the laser power density, the diameter of the laser beam, and the travel speed.

9.2.3 Laser Shock Peening

The LSP method was first developed in 1972 at the Battelle Memorial Institute. LSP is used to induce the compressive residual stress in the specimen in order to enhance the fatigue properties without adding any layer. Figure 9.4 represents the schematic representation of the LPS process, which

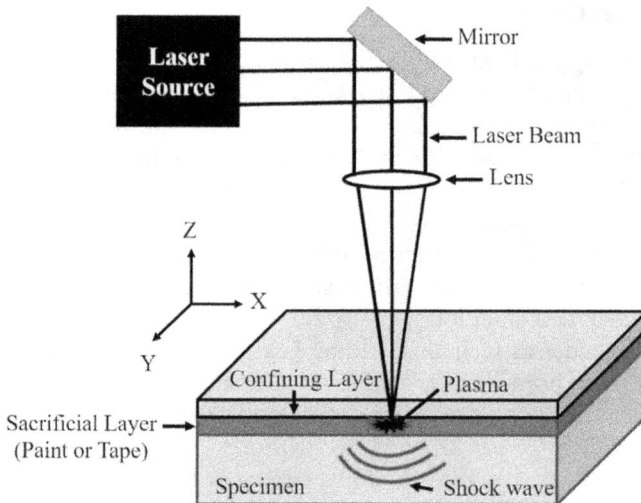

Figure 9.4 Schematic diagram of LSP process.

consist of a laser source, mirror, optics, confining layer, and the sacrificial layer. The LSP is carried out when laser pulses that are focused on the top surface bypass through the confined layer (water or air) and evaporate an ablative layer, which is often black paint or thin metallic foil. This produces hot plasma, the expansion of which drives shock waves into the material that is being peened. The confined medium increases the induced plasma pressure by obstructing the plasma expansion. It is widely used for metal components to strengthen their surface, making them less susceptible to fatigue and wear. The LSP has various advantages over traditional peening, such as higher penetration depth, noncontact process, precise protection, longer component life, and high-temperature durability. The technique finds its application in the aerospace industry for the treatment of jet engine turbine blades and in biomedical such as spinal and bone implants.

9.2.4 Laser Surface Melting

LSM, also known as laser re-melting, is a process that uses heat to melt the surface, followed by quick solidification that improves mechanical and surface properties without adding additional layers or changing the bulk material properties. The melt pool depth can be controlled using the laser power, mode of operation, scanning speed, and standoff distance. The SLM process's primary objectives are refining microstructure, dissolution of precipitates, and homogenizing composition [36]. LSM is used in a variety of materials to enhance their microhardness, wear, and corrosion properties. LSM utilized Nd: YAG or CO_2 laser as the source, mirror, and focusing lens, as shown in Figure 9.5. In order to avoid oxidation, argon (Ar) is employed as a shielding gas throughout the SLM process. Furthermore, the LSM process is also used for polishing operations by reducing the surface roughness of the specimen [37]. LSM is extensively employed for HEA coatings to enhance their microhardness, surface, wear, and corrosion properties [4,38].

9.2.5 Laser Hardening

The LH process utilizes a high-power laser beam to heat the surface and solidify it via a self-quenching method, resulting in grain refinement and higher hardness. LH has various advantages over the traditional heat treatment technique, such as control over the hardening depth, greater compatibility with a smaller component, higher accuracy, decreased distortion, and low HEZ. It enhances the hardness, wear, and corrosion resistance of the components. LH using a CO_2 laser was first applied to the diesel engine cylinder liner in 1979. The hardened cylinder liner performs better and enhances its service life by 30% post-LH [39]. A schematic diagram of the LH process is shown in Figure 9.6. The LH process is widely used in aerospace, nuclear, automobile, and construction applications. The

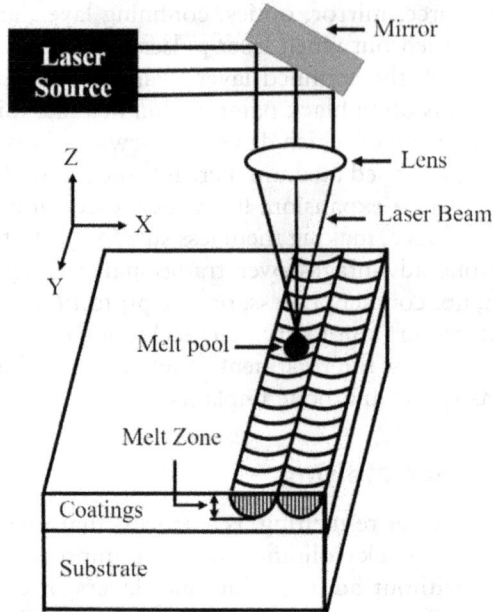

Figure 9.5 Schematic layout of laser re-melting process.

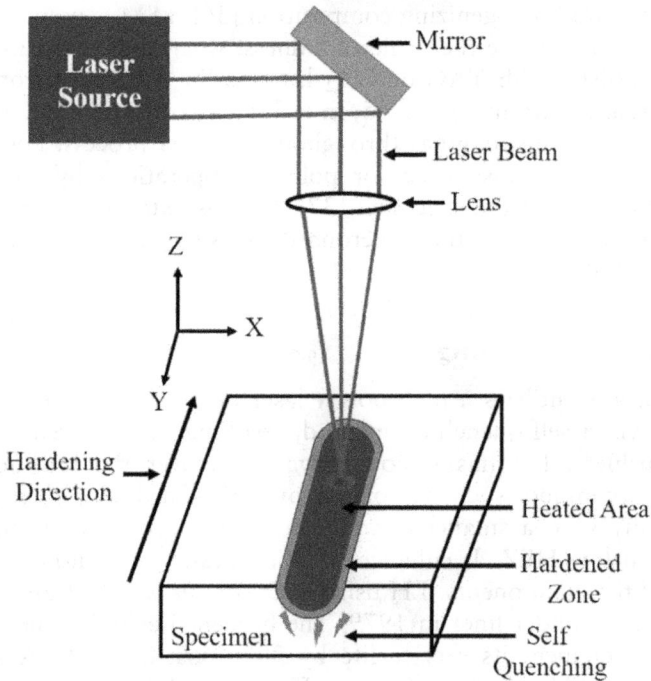

Figure 9.6 A schematic representation of the LH process.

LH process has several disadvantages, such as it only utilizes the 10% of input energy, the remaining 90% of energy reflects, higher initial cost, and higher alloy steel is difficult to process.

9.2.6 Laser Surface Alloying

The LSA process utilizes the high-power density laser beam to alloy the pre-placed powder elements on the specimen surface by adding a layer. It is widely applied for a range of materials such as Ti-based alloys, Al-based alloys, Cu-based alloys, Mg-based alloys, and Ni-based alloys to improve the microhardness, wear, and corrosion properties. The elemental surface distribution, alloy material choice, and alloy composition all affect the properties of the alloyed laser component. A schematic of the LSA process is shown in Figure 9.7. The LSA is widely used in aerospace applications, resistance welding electrodes, and automobile and electrical appliances to protect them from wear and corrosion. LSA is recently used for metallic glasses in order to enhance microhardness [40]. This process poses various advantages such as minimal distortion, strong metallurgical bonding, and tailored surface composition. However, LSA has some drawbacks, such as porosity formation, susceptibility to crack, overlapping needed, and higher initial cost.

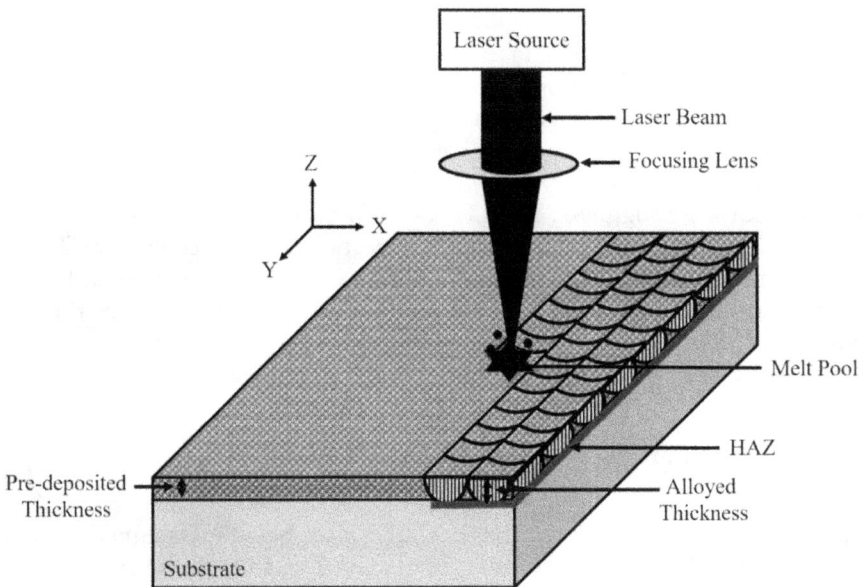

Figure 9.7 A basic layout of laser alloying process.

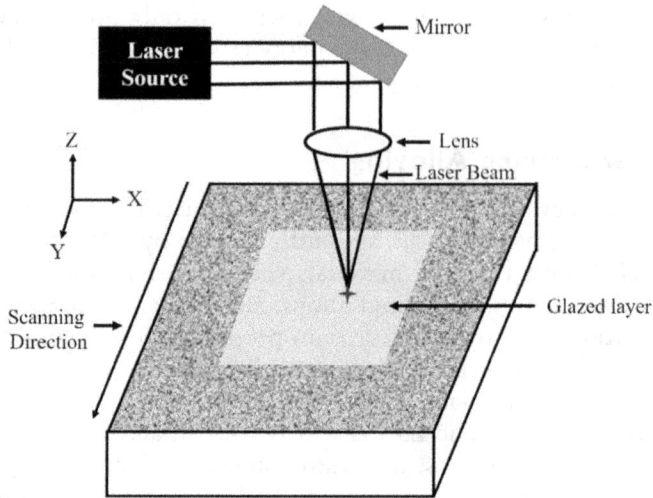

Figure 9.8 Schematic representation of LG process.

9.2.7 Laser Glazing

LG is a surface modification technique that produces a thin coating of melted material by continuously tracing a high-energy laser beam over a substrate surface. Generally, a continuous CO_2 laser source is used for LG. This process is known as LG because the material's surface looks glassy once the thin melted layer solidifies. A schematic diagram of the LG process is shown in Figure 9.8. The laser ablation during the process is controlled using the low-power density. This can be applied to both metals and ceramics materials. The glassy, amorphous, or non-crystalline surfaces may be observed post-LG. LG process has various advantages such as high heat transmission, precision, repeatability, low-energy wastage, higher density, and low surface roughness. The LG is widely used to improve the hardness, thermal barrier, wear and corrosion properties of the specimen, and mitigate the stress corrosion cracking. Recently, LG is used for thermal barrier coating and nuclear water reactor application to minimize the thermal effect and stress corrosion cracking [41,42]. Besides, it has several disadvantages, such as higher operating costs, and overlapping is needed.

9.3 LASER–MATERIAL INTERACTION

Laser–material interaction (LMI) is crucial in laser-based surface modification methods. The laser interacts with the material in four steps, as shown in Figure 9.9. LMIs are resonant and non-resonant. Resonant interactions generate photons and heat. However, when photons of a specific wavelength are absorbed and then quickly reemitted at the same wavelength, this is referred to

Laser power absorption

Laser beam heating

Laser beam melting

Convection and solidification

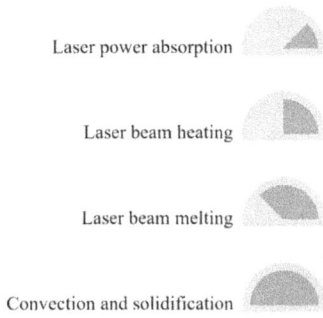

Figure 9.9 Process involved in the LMI.

non-resonant interaction. The laser power absorption by the material; absorption of laser energy is critical for laser-based surface processing, and laser irradiance should be constant. The absorptivity depends upon the material absorption coefficient, types of laser used, specimen surface roughness, laser power density, the incident angle of the laser beam, and specific mechanisms for different types of materials. When subjected to a laser beam, materials exhibit effects such as reflection, absorption, and light transmission. The laser beam interacts with the material surface at a specific amount of kinetic energy. Laser light is converted into kinetic energy in the form of free electrons rather than heat, which is absorbed by the substance. However, there are various mechanisms involved in converting absorbed laser energy into heat. The first of the three stages of the laser's power transformation into heat is the collision between excited particles, which causes their motion in space and time to be randomly shuffled. In the second stage, energy is transferred to the lattice by a variety of mechanisms, and in the last stage, heat is conducted from the high to the low-temperature zone. In order to melt the surface, the laser power, material thermal and optical properties, and the wavelength of the light all play a vital role. The laser beam melts the material when its temperature rises to the melting point. The enthalpy of melting and latent heat are the primary causes of material melting [14]. The melting of material is defined by the ratio of melt enthalpy to required melt energy; this ratio should be less than 1 in the case of metal and greater than 1 in the case of semiconductors [43]. After laser melting, the material solidifies, and crystalline or amorphous atomic structures are formed, depending upon the rate of solidification. The nucleation growth and phase transformation occur above the glass transition temperature. Convection re-solidify the material to room temperature without altering its atomic structure [43].

9.4 AREA OF APPLICATIONS

Laser-based surface modification techniques play a crucial role in improving the desired material surface properties. With the use of lasers, it is possible to

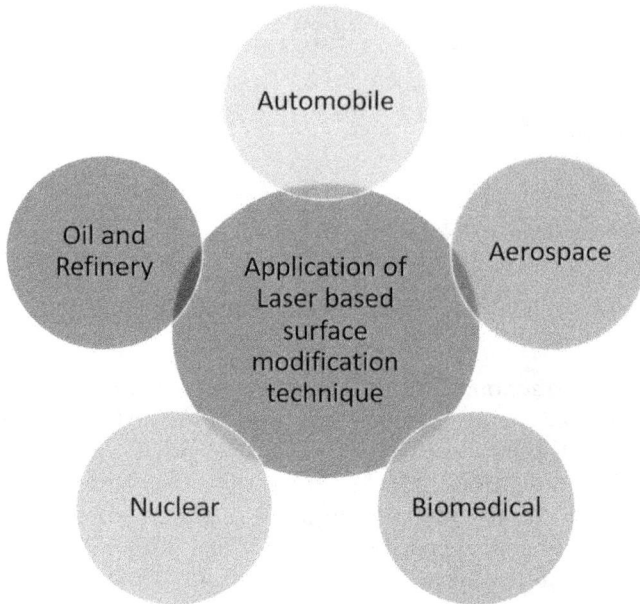

Figure 9.10 Applications of laser-based surface modification technique.

precisely direct vast quantities of energy into constrained material areas to achieve the desired properties. This energy is absorbed close to the surface of materials, changing the microstructure, crystal structure, surface chemistry, and multiscale morphology without changing the bulk material properties. It has a more remarkable ability to target specific areas of surfaces, control over the heat input, shorter processing time, noncontact process, and complex shaped specimens can also be processed. Laser-based surface modification techniques have a wide range of applications, including but not limited to those mentioned in Figure 9.10, on account of their processing capabilities. Further, this section discusses a range of applications in detail.

9.4.1 Automobile Applications

Today, energy conservation is one of the most serious concerns in scientific development. It is feasible to estimate the cost savings throughout the lifespan for the decreased fuel consumption due to the lower weight. The cost saving is around 750 Rs/kg for a typical automobile [44]. An automobile disc brake uses gray cast iron, which is heavier and consumes more power and CO_2 emission to the environment. Nonetheless, the aluminum and titanium-based alloys significantly reduced the consumption of fuel and CO_2 emission. The vehicle's reduced weight and less wear and friction lead to significant energy savings and are economically efficient. Recently, laser nitrided titanium alloy performed better than gray cast iron in terms of enhanced hardness, wear resistance, and lower weight, suggesting that it may be used as a lighter disc brake rotor [45]. Furthermore, Schubert et al. [46] used a laser-based joining process to weld Al-steel and Al-mg alloy for lightweight automotive body applications. However, LC of Al-TiC and Al-TiC-Y$_2$O$_3$ material was deposited on AZ63-Er alloy to improve the corrosion property, which is used for automobile engine components. The results indicated that the laser-cladded samples exhibit a 2.75 times higher hardness than the substrate and are free from surface imperfection [47]. The LC is also used to repair the damaged rail wheel [48]. Laser surface hardening was used on the cylinder liner, and the results revealed an improvement in the microhardness and tribological properties of the gray cast iron liner [49]. The use of laser marking technology has increased in the auto industry as a result of the development of novel techniques. It has gained attention in automobile components such as irregular surface marking, engraving, and light marking.

9.4.2 Aerospace Applications

Aerospace applications require the materials to have a higher strength-to-weight ratio, ductility, and formability. However, mechanical properties such as hardness, wear, corrosion, and fatigue resistance are the prerequisite in closer surface regions. Laser-based surface modification techniques are widely used to improve the aforementioned properties. Kumar et al. used the LSP technique to enhance the surface properties of SiC and Zirconia-reinforced Al7075 alloy for aerospace applications [50]. The results revealed an improvement in the microhardness value by 1.69 times, and surface roughness is reduced post-LSP. Hybrid laminar flow control is a promising technique to decrease fuel consumption in the aerospace industry. The reduction of drag on the trail and wins of the aircraft required micromoles in the range of 50–100 μm. Drilling such precise holes on a wide range of materials is challenging nowadays. Stephen et al. used laser drilling for an aerospace application and bored a hole that was 30 m in diameter at a rate of more than 400 holes per second in 0.8 mm thick titanium sheets [51]. In the aerospace sector, permanent marking procedures are needed for traceability and identification of components. Laser marking is a cutting-edge method in contrast to the traditional marking method. It facilitates high repeatability, noncontact process, and high scanning speed. Velotti et al. used the 30 W Nd: YAG laser to mark the titanium layer for the aerospace application [52]. Furthermore, Turner et al. used Nd: YAG laser to clean the gaseous impurities of titanium-based aero-engine components [53]. The laser re-melting was employed on the Al–1.5 wt.% Fe aerospace alloy using a Yb fiber laser to improve the corrosion resistance [54]. The findings indicate an increase in hardness and corrosion resistance post-laser re-melting due to the formation of the protective oxide film. Jagdheesh et al. produced super hydrophobic Al7075 aerospace alloy using laser patterning; the specimen fabricated at 100 and 150 μm pitch shows superior hydrophobic properties [55].

9.4.3 Biomedical Applications

The laser surface modification techniques for biomedical applications have been widely used in recent years in order to improve biocompatibility, corrosion, fatigue, and wear resistance. The materials used in biomedical implants must fulfill the stringent standards of ISO 13356:2008. In particular, LST has drawn more attention as a means of accurate surface topography, adjustable surface texturing with complicated geometries, decreased heat-affected zone, applicable for a wide range of materials, and does not effect the bulk properties. Recently, LST of nano and micron size features was performed at Ti-based bioimplant material with improved proliferation and tissue cell wetting properties, suitable for biomedical application [56–59]. Furthermore, selective laser melting was used to fabricate microporous Ti–6Al–4V alloy with enhanced osteoblastic cell compatibility, adhesion, proliferation, and bioactivity [60,61]. In addition, Hammadi used Nd: YAG laser to produce the microchannel for microfluidic devices [62]. LC is also used to improve the biocompatibility of various materials for biomedical applications [63–65].

9.4.4 Nuclear Applications

The research and development in nuclear power generation have gained attention in the past decade due to increasing electricity demand and regulations for reliable and clean energy. Generally, the nuclear reactor operates at high temperatures, and a harsh environment leads to surface degradation. In order to enhance the hardness and wear resistance of the reactor materials, LSP was used [66]. Furthermore, LSP is also used to improve the stress corrosion cracking in nuclear power plants by inducing compressive residual stress [67]. In addition, LC is used for the reclamation of Inconel heat exchanger tubes in nuclear applications [68]. Pacquentin et al. employed LSM to limit nickel emission from pressurized water nuclear reactors. The results exhibit the lower release of Ni-rich in chromium, which protects from corrosion [69]. In order to avoid stress corrosion cracking in nuclear applications, Stutzman et al. adopted the LG technique. The findings revealed that a combination of slow traversal speeds and high laser power creates deep fusion zones and disseminates the pre-existing creak, preventing stress corrosion cracking [42].

9.5 HIGH-ENTROPY ALLOY

HEAs are new emerging materials which contain 5 at % to 35 at % in equiatomic or near equiatomic composition. These alloys' remarkable characteristics are a direct result of their novel design idea. As a result, this new area of study has gained a lot of attention in the past decade. In their liquid or random solid–solution phases, HEAs have far greater mixing entropies than conventional alloys. The formation of solid solutions, both disordered and partly ordered phase, is formed due to the high mixing entropy [1]. In recent years, we have seen many studies devoted to the development of criteria that use characteristics like mixing entropy/configurational entropy (ΔS_{conf}), enthalpy of mixing (ΔH_{mix}), gibes free energy of mixing (ΔG_{mix}), atomic size difference (δ), valence electron concentration (VEC), and dimension less parameter (Ω) to predict the phase formation in the alloy. The configurational entropy of the alloy system can be calculated as follows:

$$\Delta S_{conf} = -R \sum_{i=1}^{N} m_i \ln m_i \tag{9.1}$$

where R corresponds to novel gas constant 8.134 J/K mol, N is the number of elements present in the system, and m_i is the composition of the ith element. If $\Delta S_{conf} \leq R$, then it is classified as a conventional alloy; if $R < \Delta S_{conf} \leq 1.5R$, then it is a medium entropy alloy; and if $\Delta S_{conf} \geq 1.5R$, then it is called an HEA, as shown in Figure 9.11. Furthermore, mixing enthalpy of the alloy system can be calculated as:

$$\Delta H_{mix} = 4 \sum_{i=1, J \neq 1}^{N} \Delta H_{ij}^{mix} m_i m_j \tag{9.2}$$

where ΔH_{ij}^{mix} is the mixing enthalpy of the ith and jth elements, and m_i and m_j are the compositions of the ith and jth elements, respectively.

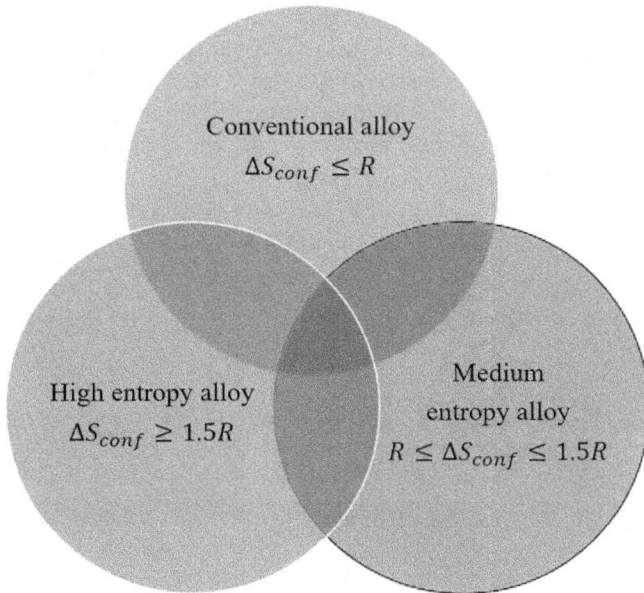

Figure 9.11 Classification of alloys based on configurational entropy.

Gibbs free energy of mixing can be calculated as:

$$\Delta G_{mix} = \Delta H_{mix} - T\Delta S_{mix} \qquad (9.3)$$

where ΔH_{mix} is the mixing enthalpy, T is the temperature, and S_{mix} is the mixing entropy. If the number of elements increases in the alloy system, then it leads to higher mixing entropy and hence lower Gibbs free energy of mixing. Despite the fact that an alloy may be in a variety of states, the second rule of thermodynamics stipulates that the equilibrium state is the one with the lowest free energy of mixing.

Another criterion to predict the phase formation in HEAs is atomic size difference. The atomic size difference can be calculated as follows:

$$\delta = 100\sqrt{\sum_{i=1}^{N} m_i\left(\frac{r_i}{\bar{r}}\right)^2} \qquad (9.4)$$

where m_i and r_i are the compositions and atomic radius of the ith element, respectively, whereas $\bar{r}$ is the average atomic radius of all the elements present in the alloy system. It can be calculated as: $\bar{r} = \sum_{i=1}^{N} m_i r_i$. In the past decade, many researchers have given various criteria for the formation of the solid–solution phase. Zhang et al. [70] proposed criteria for the formation of a solid–solution phase if 12 J/K mol $\leq \Delta S_{mix} \leq$ 17.5 J/K mol, -20 kJ/mol $\leq \Delta H_{mix} \leq$ 5 kJ/mol, and $\delta \leq 6.4\%$. Furthermore, the criteria for the formation of the disordered phase are given as $\delta \geq 4.6\%$ and $\Delta H_{mix} \geq -15$ kJ/mol. In addition, Sheng and Liu [71] proposed the criteria for solid–solution formation if $\delta \leq 6.6\%$ and -11.6 kJ/mol $\leq \Delta H_{mix} \leq$ 3.2 kJ/mol. In the case of a negative enthalpy of mixing, $\Delta H_{mix} > -11.6$ and $\delta > 6.6\%$, an amorphous phase will be formed. Furthermore, Ren et al. [72] suggested a criterion for the formation of the solid–solution phase, if $\delta \leq 2.77\%$ and $\Delta H_{mix} \geq -28.8$ kJ/mol.

The phase formation in the HEA can also be predicted using the VEC, and it can be calculated using equation (9.5)

$$\text{VEC} = \sum_{i=1}^{N} m_i(\text{VEC})_i \qquad (9.5)$$

where m_i and VEC_i are the composition and VEC of the ith element, respectively. N represents the number of elements in the alloy system. Gao et al. [73] used VEC criteria for the prediction of BCC and FCC phase formation in HEAs. The BCC phase forms at a VEC < 6.87; however, the FCC phase forms at a VEC ≥ 8.

Afterwards, a dimensionless parameter Ω was introduced by Zhang et al. [74] to predict the phase formation in the HEAs given as

$$\Omega = \left| \frac{T_m \, \Delta S_{mix}}{\Delta H_{mix}} \right|, \tag{9.6}$$

where ΔS_{mix} is the mixing entropy, ΔH_{mix} is the mixing enthalpy, and T_m is the theoretical melting temperature, and it can be calculated as $T_m = \sum_{i=1}^{N} m_i (T_m)_i$, where N is the number of elements in the alloy system, and m_i and $(T_m)_i$ are the composition and melting temperature of the ith element, respectively. According to these criteria, the solid solution will form if $\Omega \geq 1.1$ and $\delta \leq 6.6\%$.

HEAs have unique features due to the composition of the multiprincipal elements. HEAs pose higher hardness, remarkable high-temperature strength, structural stability, superior wear, corrosion, and oxidation resistance. Since some of these characteristics are not present in traditional alloys, HEAs are attractive in various industries. Its application range is considerably broadened by the fact that it exhibits higher temperature stability [4].

9.6 EXPERIMENTAL PROCEDURE

Prior to deposition, a mild steel plate with the dimensions $100 \times 100 \times 5$ mm was sectioned using wire electrical discharge machining and cleaned the plate with an ethanol bath. The substrate's cleanliness, surface roughness, and powder morphology all play a major role in adhesion and bonding between substrate and coating. All of the powder particle elements were properly blended in the proper composition before deposition. The detailed chemical composition of the alloy is given in our previous work [3]. To increase the effectiveness of the atmospheric plasma spray deposition, the powder particle size and shape are crucial. In order to improve the deposition efficiency, the particle shape is chosen as spherical and 10–40 μm in size. The spherical shape powder was preferred for deposition due to its high flow ability, and splatted on the substrate homogeneously leads to reduced porosity. As a consequence of the high impact speed, powder particles with a size of less than 10 μm are blown away, while particles with a size of more than 40 μm do not have sufficient homogeneity and diffusion in the sample that is deposited. The pre-mixed alloy powder is preheated before the deposition to remove the moisture content. The schematic plasma gun used during the atmospheric plasma sprayed (APS) deposition process is shown in Figure 9.12. The argon (Ar) is used as primary plasma gas, and hydrogen is selected as secondary plasma gas. Argon is used as a primary plasma gas because it is an inert gas and prevents oxidation during the deposition process. In order to produce a spark, a power source must be connected between the anode and the cathode. Furthermore, argon gas is ionized into plasma when it comes into contact with a spark. The plasma jet flows through a heat and shock resistance convergent nozzle. After being

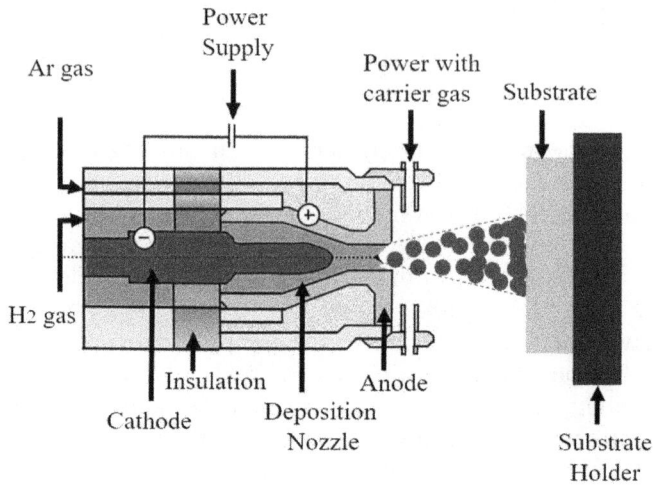

Figure 9.12 A schematic representation of the atmospheric plasma deposition process.

pre-mixed, $CoCrFeNiW_{0.3}$ alloy powder with proper composition was loaded into the hopper, where it was then driven into the plasma jet via the use of argon as the carrier gas. The pre-mixed powder particle absorbed the heat during the inflight and impacted at supersonic speed on the substrate, got flattened and solidify.

$CoCrFeNiW_{0.3}$ alloy coating was deposited at a mild steel substrate, and the deposited coating with the substrate is shown in Figure 9.13. The visual inspection confirms the strong metallurgical bonding between the substrate and alloy. There are no visible cracks or pores observed in the deposited sample.

Furthermore, the LSM was performed on the as-deposited sample in order to improve the surface and mechanical properties. Prior to the LSM, the coated alloy samples were cleaned using ethanol. The schematic setup of LSM is shown in Figure 9.14. The setup comprised of the laser source, mirror, focusing lens, and substrate holder. The Nd: YAG (neodymium-doped yttrium aluminum garnet) is used as a laser source. Nd: YAG laser is preferred over other lasers because it consumes lower power, higher gain, and improved efficiency. Generally, the Nd: YAG laser can be operated in the pulsed and continuous mode. In the current study, the pulsed mode laser opted over the continuous mode because the pulsed mode laser has a higher peak power and greater penetration depth. Nd: YAG laser consists of three basic elements: active medium, energy source, and optical resonator. Nd: YAG is used as the active medium. The energy source is used to excite the electron of neodymium from a lower energy state to a higher energy state, leading to population inversion. Various energy sources are used, such as diode lasers, arc lamps, and flash lamps, to pump the active medium. The active medium is placed between the two mirrors, called a

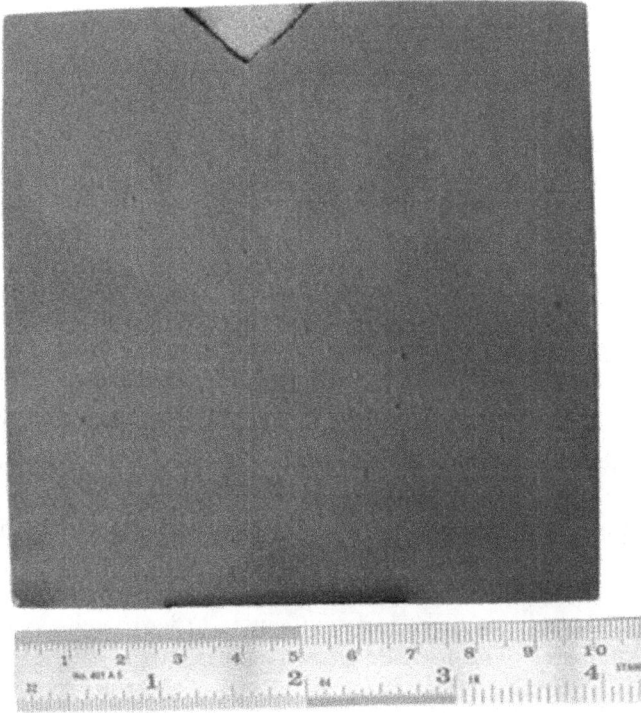

Figure 9.13 APS CoCrFeNiW$_{0.3}$ HEA.

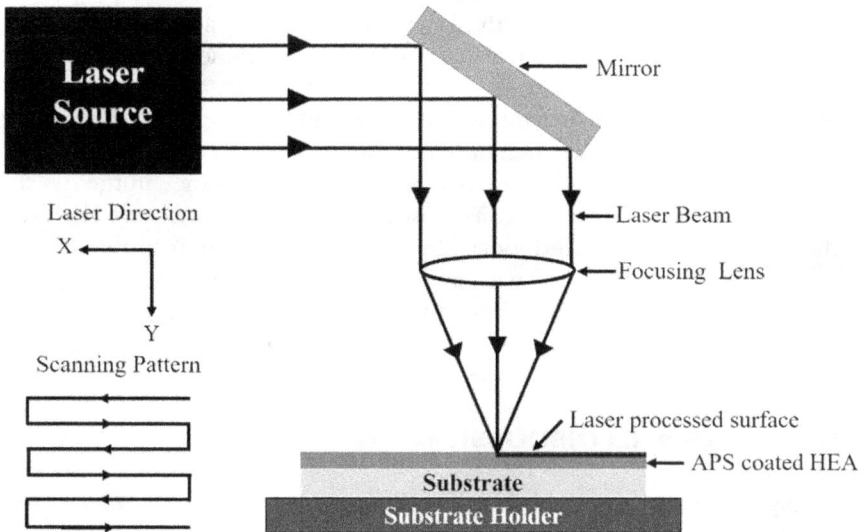

Figure 9.14 Schematic of LSM process.

Table 9.1 Optimized process parameters used for the LSM [14]

S.No.	Parameter	Unit	Value
1.	Spot diameter	mm	2
2.	Power	W	210
3.	Frequency	Hz	10
4.	Scanning speed	mm/min	2.8
5.	Hatch spacing	mm	4
6.	Wavelength	λ	532
7.	Pulse duration	ns	10
8.	Overlap	%	90

resonator. One mirror is a reflecting mirror, which bounces all the laser light back, while the other is a partially reflecting mirror, which allows some of the laser light to pass through to produce a laser beam. The light from the active medium is reflected back and forth before leaving the resonator. Q-switch mode was used to introduce the loss and produce the short-pulsed laser. This short-pulsed laser beam is reflected through the mirror and passes through the focusing lens in order to achieve the desired laser spot diameter. The focused laser beam impinges the specimen; the laser energy is absorbed. The absorbed energy is converted into heat. The absorbed energy depends upon the coefficient of heat absorption. The absorbed heat leads to surface melting and solidification.

The laser process parameter is crucial in achieving the desired properties. In order to avoid laser ablation during the LMI process, the LSM process parameters, such as scanning rate, laser power density, and hatch spacing, can be optimized. The optimized laser process parameters are given in Table 9.1. Laser scanning was performed using a hatch pattern on the APS-coated CoCrFeNiTi HEA sample. A finite-element scanning microscope (FESEM) was used to investigate the microstructural change pre- and post-LSM. Archimedes' principle was used to calculate the change in the density pre- and post-LSM. The optical microscope was used to measure the areal surface roughness pre- and post-LSM in order to confirm the surface quality. The microhardness was checked using Vickers's microhardness tester before and after the LSM process. The observed results are discussed in the subsequent section.

9.7 RESULTS AND DISCUSSION

Scanning electron spectroscopy was used to analyze the $CoCrFeNW_{0.3}$ alloy coating pre- and post-LSM. $CoCrFeNW_{0.3}$ alloy coating exhibits porosity and a few partially melted particles at the surface pre-LSM, as shown in Figure 9.15(a). Less dwell time during the powder particle injection in APS

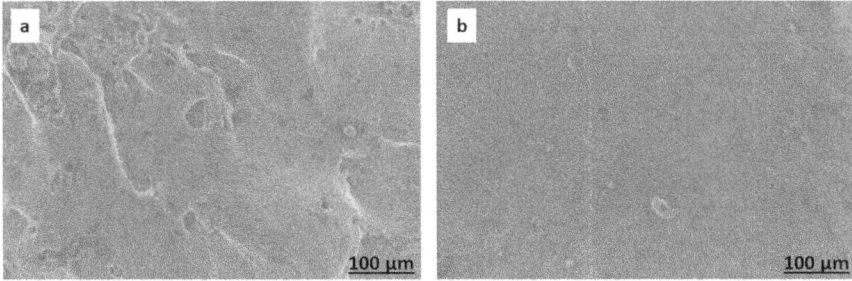

Figure 9.15 FESEM image of CoCrFeNiW0.3 HEA: (a) as-sprayed; (b) post-LSM.

may be responsible for the partially melted particles. Moreover, due to the impact and splat of the powder particles, porosity is also seen at the sample's surface. However, $CoCrFeNW_{0.3}$ alloy coating shows a lower porosity and improved surface finish, as shown in Figure 9.15(b). The decreased porosity and improved surface finish are attributed due to localized heating and solidification. High-density pulsed laser has the ability to melt any high thermal absorption coefficient metals. The alloy's density was measured pre- and post-LSM using the Archimedes principle. The laser surface melted alloy exhibits an 11.56% enhanced density in contrast to the as-sprayed samples. The optical profilometer is used to measure the areal surface roughness (S_a) pre- and post-LSM. The $CoCrFeNiW_{0.3}$ alloy coating shows a significantly decreased in areal surface roughness by ~30% post-LSM. The decreased surface roughness post-LSM attributed due to localized heating and rapid solidifications leads to lower surface defects and pores present on the surface.

The LSM alloy's surface roughness is reduced, which results in improved the alloy's mechanical properties. To investigate how the mechanical characteristics improved throughout the laser melting process, a micro-hardness test was conducted pre- and post-LSM process. The samples were polished using the SiC paper of various grits size according to the standard metallographic method prior to the microhardness test. The Vickers hardness indenter was used under the 200 gf load and 30s dwell time to ensure the proper impression on the sample. To determine the sample's average hardness value, the indenter impingement was applied ten times without overlapping. A microhardness test revealed that the laser surface melted specimen exhibits an improved hardness value by ~20%, as shown in Figure 9.16. The improved hardness value is attributed due to enhanced surface density and rapid cooling during the LSM.

9.8 CONCLUSIONS

LSM is a versatile method for modifying surfaces using a laser that does not involve changing the underlying chemical content of the sample.

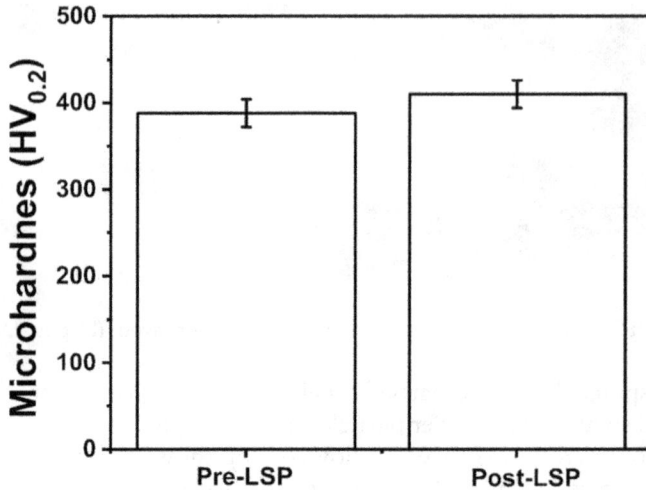

Figure 9.16 Microhardness graph of CoCrFeNiW alloy pre- and post-LSM.

The morphology, microstructure, roughness, and microhardness of a material's surface may all be enhanced with this method. In this chapter, the APS-coated HEA undergoes pulsed LSM to increase its surface attributes, such as surface roughness and microhardness, for use in wear and corrosion resistance applications. Therefore, this investigation has the following conclusions.

- Laser surface melted $CoCrFeNiW_{0.3}$ alloy exhibits lower surface roughness and better surface finish.
- $CoCrFeNiW_{0.3}$ alloy displays the improved surface hardness value post-LSM.
- Laser surface melted $CoCrFeNiW_{0.3}$ alloy is suitable for wear and fatigue-resistant applications because of the enhanced surface quality and hardness.

9.9 FUTURE SCOPE

The presented technique elucidates how the impact of LSM effectively introduces new practices to HEAs. The behavior of HEAs after being subjected to traditional furnace-based heat treatment may be examined with that of HEAs subjected to laser-based treatments in the future. The aforementioned findings will provide some insight into how HEAs respond to conventional and laser-based unconventional synthesis techniques.

ACKNOWLEDGMENTS

The authors thank the financial support provided by the Indian Space Research Organization (ISRO) for providing the financial support to carry out the experiments through the Respond Scheme (Project Sanction no. ISRO/RES/3/844/19-20).

REFERENCES

[1] Yeh, J. W., Chen, S. K., Gan, J. Y., Lin, S. J., Chin, T. S., Shun, T. T., Tsau, C. H., and Chang, S. Y. Formation of simple crystal structures in Cu-Co-Ni-Cr-Al-Fe-Ti-V alloys with multiprincipal metallic elements. *Metall. Mater. Trans. A – Phys. Metall. Mater. Sci.*, 35A, 2004, pp. 2533–2536.

[2] Yeh, J. W., Chen, S. K., Lin, S. J., Gan, J. Y., Chin, T. S., Shun, T. T., Tsau, C. H., and Chang, S. Y. Nanostructured high-entropy alloys with multiple principal elements: Novel alloy design concepts and outcomes. *Adv. Eng. Mater.*, 6, 2004, pp. 299–303.

[3] Kumar, H., Bhaduri, G. A., Manikandan, S. G. K., et al. Microstructural characterization and tribological properties of atmospheric plasma sprayed high entropy alloy coatings. *J. Therm. Spray Technol.*, 31, 2022, pp. 1956–1974.

[4] Kumar, H., Bhaduri, G. A., Manikandan, S. G. K., et al. Influence of annealing on microstructure and tribological properties of AlCoCrFeNiTi high entropy alloy based coating. *Met. Mater. Int.*, 2022, 10.1007/s12540-022-01264-y

[5] Zhicheng, L. and Dejun, K. Structural and electrochemical corrosion properties of plasma-sprayed CoCrFeNiMo HEA coating in corrosive solutions. *Corros. Eng. Sci. Technol.*, 2022, DOI: 10.1080/1478422X.2022.2120945

[6] Wang, Z.-Q., Shi, Y., and Wang, X.-R. Microstructure, thermodynamics and compressive properties of Al0.5CoCrCuFeNi-X (X=V,Si,VSi) high-entropy alloys. *Mater. Sci. Technol.*, 2022, DOI: 10.1080/02670836.2022.2092999

[7] Mahaffey, J., Vackel, A., Whetten, S., et al. Structure evolution and corrosion performance of CoCrFeMnNi high entropy alloy coatings produced via plasma spray and cold spray. *J. Therm. Spray Technol.*, 31, 2022, pp. 1143–1154, 10.1007/s11666-022-01373-5

[8] Abhijith, N. V., Kumar, D., and Kalyansundaram, D. Development of single-stage TiNbMoMnFe high-entropy alloy coating on 304L stainless steel using HVOF thermal spray. *J. Therm. Spray Technol.*, 31, 2022, pp. 1032–1044, 10.1007/s11666-021-01294-9

[9] Yao, X., Peng, K., Chen, X., Jiang, F., Wang, K., and Wang, Q. Microstructure and mechanical properties of dual wire-arc additive manufactured Al-Co-Cr-Fe-Ni high entropy alloy. *Mater. Lett.*, 326, 2022, p. 132928, 10.1016/j.matlet.2022.132928

[10] Ren, B., Lv, S.-J., Zhao, R.-F., Liu, Z.-X., and Guan, S.-K. Effect of sputtering parameters on (AlCrMnMoNiZr)N films. *Surf. Eng.*, 30, 2014, pp. 152–158, DOI: 10.1179/1743294413Y.0000000226

[11] Yue, T. M. and Zhang, H. Laser cladding of FeCoNiCrAlCuxSi0·5 high entropy alloys on AZ31 Mg alloy substrates. *Mater. Res. Innov.*, 18, 2014, pp. 624–628, DOI: 10.1179/1432891714Z.000000000530

[12] Bande, H., L'Espérance, G., Islam, M. U., and Koul, A. K. Laser surface hardening of AISI 01 tool steel and its microstructure. *Mater. Sci. Technol.*, 7, 1991, pp. 452–457, DOI: 10.1179/mst.1991.7.5.452

[13] Dong, J. L., Wu, X. Q., and Huang, C. G. Mechanical behavior of thin CoCrFeNi high-entropy alloy sheet under laser shock peening. *Intermetallics*, 144, 2022, p. 107529, 10.1016/j.intermet.2022.107529

[14] Kumar, H., Kumar, C., Manikandan, S. G. K., Kamaraj, M., and Shiva, S. Laser re-melting of atmospheric plasma sprayed high entropy alloy. In: Radhakrishnan, J., Pathak, S. (eds.) *Advanced engineering of materials through lasers. Advances in material research and technology.* Springer, Cham, 2022, 10.1007/978-3-031-03830-3_5

[15] Chi, Y., Meng, X., Dou, J., Yu, H., and Chen, C. Laser alloying with Fe–B4C–Ti on AA6061 for improved wear resistance. *Surf. Eng.*, 37:12, 2021, pp. 1503–1513, DOI: 10.1080/02670844.2021.1916862

[16] Li, M. X., Leonard, H. R., Rommel, S., Hung, C., Watson, T. J., Policandriotes, T., Hebert, R. J., and Aindow, M. Use of laser glazing to evaluate an Al-Cr-Co-Mn-Zr alloy containing icosahedral quasicrystalline dispersoids as a candidate material for additive manufacturing. *Addit. Manuf.*, 59, 2022, p. 103114, 10.1016/j.addma.2022.103114

[17] Lin-ting, W., Xin-yi, Z., Li-na, Z., Jia-jie, K. Zhi-qiang, F., Ding-shun, S., and Run-jie, L. The influence of laser surface texture parameters on the hydrophobic properties of Fe-based amorphous coatings. *J. Non Cryst. Solids*, 593, 2022, p. 121771, 10.1016/j.jnoncrysol.2022.121771

[18] Liu, R., Yu, J., Yang, Q., Wu, J., and Gao, S. Mechanical property and strengthening mechanism on DH36 marine steel after laser surface melting. *Surf. Coat. Technol.*, 425, 2021, p. 127733, 10.1016/j.surfcoat.2021.127733

[19] Ardila-Rodríguez, L. A., Menezes, B. R. C., Pereira, L. A., Takahashi, R. J., Oliveira, A. C., and Travessa, D. N. Surface modification of aluminum alloys with carbon nanotubes by laser surface melting. *Surf. Coat. Technol.*, 377, 2019, p. 124930, 10.1016/j.surfcoat.2019.124930

[20] Jiao, Y., Brousseau, E., Kosai, K., Lunt, A. J. G., Yan, J., Han, Q., Zhu, H., Bigot, S., and He, W. Softening and hardening on a Zr-based bulk metallic glass induced by nanosecond laser surface melting. *Mater. Sci. Eng. A*, 803, 2021, p. 140497, 10.1016/j.msea.2020.140497

[21] Liu, G., Du, D., Wang, K., Pu, Z., and Chang, B. Epitaxial growth behavior and stray grains formation mechanism during laser surface re-melting of directionally solidified nickel-based superalloys. *J. Alloys Compd.*, 853, 2021, p. 157325, 10.1016/j.jallcom.2020.157325

[22] Zhang, L., Yu, G., Li, S., He, X., Xie, X., Xia, C., Ning, W., and Zheng, C. The effect of laser surface melting on grain refinement of phase separated Cu-Cr alloy. *Opt. Laser Technol.*, 119, 2019, p. 105577, 10.1016/j.optlastec.2019.105577

[23] Rakesh, K. R., Bontha, S., Ramesh, M. R., Das, M., and Balla, V. K. Laser surface melting of Mg-Zn-Dy alloy for better wettability and corrosion resistance for biodegradable implant applications. *Appl. Surf. Sci.*, 480, 2019, pp. 70–82, 10.1016/j.apsusc.2019.02.167

[24] Sassmannshausen, A., Brenner, A., and Finger, J. Ultrashort pulse laser polishing by continuous surface melting. *J. Mater. Process. Technol.*, 293, 2021, p. 117058, 10.1016/j.jmatprotec.2021.117058

[25] Chen, C., Zhang, H., Hu, S., Wei, R., Wang, T., Cheng, Y., Zhang, T., Shi, N., Li, F., Guan, S., and Jiang, J. Influences of laser surface melting on microstructure, mechanical properties and corrosion resistance of dual-phase Cr–Fe–Co–Ni–Al high entropy alloys. *J. Alloys Compd.*, 826, 2020, p. 154100, 10.1016/j.jallcom.2020.154100

[26] Chakraborty, P., Kumar, S., and Tewari, R. Effect of laser re-melting on the microstructure of high entropy alloys. *Mater. Lett.*, 324, 2022, p. 132669, 10.1016/j.matlet.2022.132669

[27] Lyu, P., Gao, Q., You, X., Peng, T., Yuan, H., Guan, Q., Cai, J., Liu, H., Liu, X., and Zhang, C. High temperature oxidation enhanced by laser re-melted of CoCrFeNiAlx (x = 0.1, 0.5 and 1) high-entropy alloys. *J. Alloys Compd.*, 915, 2022, p. 165297, 10.1016/j.jallcom.2022.165297

[28] Jin, B., Zhang, N., Guan, S., Zhang, Y., and Li, D. Microstructure and properties of laser re-melting FeCoCrNiAl0.5Six high-entropy alloy coatings. *Surf. Coat. Technol.*, 349, 2018, pp. 867–873, 10.1016/j.surfcoat.2018.06.032

[29] Erdogan, A., Döleker, K. M., and Zeytin, S. Effect of laser re-melting on electric current assistive sintered CoCrFeNiAlxTiy high entropy alloys: Formation, micro-hardness and wear behaviors. *Surf. Coat. Technol.*, 399, 2020, p. 126179, 10.1016/j.surfcoat.2020.126179

[30] Cai, Y., Cui, Y., Zhu, L., Tian, R., Geng, K., Li, H., and Han, J. Enhancing the (FeMnCrNiCo + TiC) cladding layer by in-situ laser re-melting. *Surf. Eng.*, 37:12, 2021, pp. 1496–1502, DOI: 10.1080/02670844.2020.1868651

[31] Chai, L., Xiang, K., Xia, J., Fallah, V., Murty, K. L., Yao, Z., and Gan, B. Effects of pulsed laser surface treatments on microstructural characteristics and hardness of CrCoNi medium-entropy alloy. *Philos. Mag.*, 99:24, 2019, pp. 3015–3031, DOI: 10.1080/14786435.2019.1649499

[32] Smurov, I., Uglov, A., Krivonogov, Y., et al. Pulsed laser treatment of plasma-sprayed thermal barrier coatings: Effect of pulse duration and energy input. *J. Mater. Sci.*, 27, 1992, pp. 4523–4530, 10.1007/BF00541589

[33] Novak, T. G., Vora, H. D., Mishra, R. S., et al. Synthesis of Al0.5CoCrCuFeNi and Al0.5CoCrFeMnNi high-entropy alloys by laser melting. *Metall. Mater. Trans. B*, 45, 2014, pp. 1603–1607, 10.1007/s11663-014-0170-4

[34] Zhang, H., Pan, Y., He, Y. Z., et al. Application prospects and micro-structural features in laser-induced rapidly solidified high-entropy alloys. *JOM*, 66, 2014, pp. 2057–2066, 10.1007/s11837-014-1036-6

[35] Lin, D., Zhang, N., He, B., et al. Influence of laser re-melting and vacuum heat treatment on plasma-sprayed FeCoCrNiAl alloy coatings. *J. Iron Steel Res. Int.*, 24, 2017, pp. 1199–1205, 10.1016/S1006-706X(18)30018-9

[36] Narayanan, T. S. N. S., Park, I.-S., and Lee, M.-H. Surface modification of magnesium and its alloys for biomedical applications: Opportunities and challenges. In: Narayanan, T. S. N. S., Park, I.-S., Lee, M.-H. (eds.) *Surface modification of magnesium and its alloys for biomedical applications*. Woodhead Publishing, 2015, pp. 29–87, 10.1016/B978-1-78242-077-4.00002-4

[37] Perry, T. L., Werschmoeller, D., Li, X., Pfefferkorn, F. E., and Duffie, N. A. Micromelting for laser micro polishing of meso/micro metallic components. Proceedings of the ASME 2007 International Manufacturing Science and Engineering Conference. Atlanta, Georgia, USA, 15–18, 2007, pp. 363–369, 10.1115/MSEC2007-31173

[38] Chen, C., Yuan, S., Chen, J., Wang, W., Zhang, W., Wei, R., Wang, T., Zhang, T., Guan, S., and Li, F. A Co-free Cr-Fe-Ni-Al-Si high entropy alloy with outstanding corrosion resistance and high hardness fabricated by laser surface melting. *Mater. Lett.*, 314, 2022, p. 131882, 10.1016/j.matlet.2022.131882

[39] Ready, J. F. Applications for surface treatment. In: Ready, J. F. (ed.) *Industrial applications of lasers* (Second Edition), Academic Press, 1997, pp. 373–383, 10.1016/B978-012583961-7/50017-X

[40] Qian, Y., Zhang, D., Hong, J., Zhang, L., Jiang, M., Huang, H., and Yan, J. Microstructure and mechanical properties of SiC particle reinforced Zr-based metallic glass surface composite layers produced by laser alloying. *Surf. Coat. Technol.*, 446, 2022, p. 128784, 10.1016/j.surfcoat.2022.128784

[41] Dhineshkumar, S. R., Duraiselvam, M., Natarajan, S., Panwar, S. S., Jana, T., and Khan, M. A. Effect of laser glazing on the thermo-mechanical properties of plasma-sprayed LaTi2Al9O19 thermal barrier coatings. *Mater. Manuf. Process.*, 32:14, 2016, pp. 1573–1580.

[42] Stutzman, A. M., Rai, A. K., Alexandreanu, B., Albert, P. E., Sun, E. J., Schwartz, M. L., Reutzel, E. W., Tressler, J. F., Medill, T. P., and Wolfe, D. E. Laser glazing of cold sprayed coatings for the mitigation of stress corrosion cracking in light water reactor (LWR) applications. *Surf. Coat. Technol.*, 386, 2020, p. 125429, 10.1016/j.surfcoat.2020.125429

[43] Von Allmen, M. and Blatter, A. Melting and solidification. In: *Laser-beam interactions with materials. Springer series in materials science*, vol 2. Springer, Berlin, Heidelberg, 1995, 10.1007/978-3-642-57813-7_4

[44] Sepold, G., Schubert, E., Franz, T., and Klassen, M. Laserstrahlschweißen von Leichtbaukonstruktionen, In: *Fortschritte bei der Konstruktion und Berechnung geschweißter Bauteile, Braunschweig, DVS Berichte Bd. 187.* DVS-Verlag, Düsseldorf, 1997, pp. 65–69.

[45] Duraiselvam, M., Valarmathi, A., Shariff, S. M., and Padmanabham, G. Laser surface nitrided Ti–6Al–4V for light weight automobile disk brake rotor application. *Wear*, 309, 2014, pp. 269–274, 10.1016/j.wear.2013.11.025

[46] Schubert, E., Klassen, M., Zerner, I., Walz, C., and Sepold, G. Light-weight structures produced by laser beam joining for future applications in automobile and aerospace industry. *J. Mater. Process. Technol.*, 115, 2001, pp. 2–8, 10.1016/S0924-0136(01)00756-7

[47] Bu, R., Jin, A., Sun, Q., Zan, W., and He, R. Study on laser cladding and properties of AZ63-Er alloy for automobile engine. *J. Mater. Res. Technol.*, 9, 2020, pp. 5154–5160, 10.1016/j.jmrt.2020.03.032

[48] Zhu, Y., Yang, Y., Mu, X., Wang, W., Yao, Z., and Yang, H. Study on wear and RCF performance of repaired damage railway wheels: Assessing laser cladding to repair local defects on wheels. *Wear*, 430–431, 2019, pp. 126–136, 10.1016/j.wear.2019.04.028

[49] Xiu-bo L. I. U., Gang Y. U., Jian G. U. O., Shang, Q.-y., Zhang, Z.-g., and Gu, Y.-j. Analysis of laser surface hardened layers of automobile engine cylinder liner. *J. Iron Steel Res. Int.*, 14, 2007, pp. 42–46, 10.1016/S1006-706X(07)60010-7

[50] Kumar, G. R., Rajyalakshmi, G., and Manupati, V. K. Surface micro patterning of aluminium reinforced composite for aerospace applications through high energy pulse laser peening. *Mater. Today: Proc.*, 5, 2018, pp. 6963–6972, 10.1016/j.matpr.2017.11.359

[51] Stephen, A., Schrauf, G., Mehrafsun, S., and Vollertsen, F. High speed laser micro drilling for aerospace applications. *Procedia CIRP*, 24, 2014, pp. 130–133, 10.1016/j.procir.2014.08.002

[52] Velotti, C., Astarita, A., Leone, C., Genna, S., Minutolo, F. M. C., and Squillace, A. Laser marking of titanium coating for aerospace applications. *Procedia CIRP*, 41, 2016, pp. 975–980, 10.1016/j.procir.2016.01.006

[53] Turner, M. W., Schmidt, M. J. J., and Li, L. Preliminary study into the effects of YAG laser processing of titanium 6Al–4V alloy for potential aerospace component cleaning application. *Appl. Surf. Sci.*, 247, 2005, pp. 623–630, 10.1016/j.apsusc.2005.01.097

[54] Pariona, M. M., Teleginski, V., dos Santos, K., Machado, S., Zara, A. J., Zurba, N. K., and Riva, R. Yb-fiber laser beam effects on the surface modification of Al–Fe aerospace alloy obtaining weld filet structures, low fine porosity and corrosion resistance. *Surf. Coat. Technol.*, 206, 2012, pp. 2293–2301, 10.1016/j.surfcoat.2011.10.007

[55] Jagdheesh, R., Hauschwitz, P., Mužík, J., Brajer, J., Rostohar, D., Jiříček, P., Kopeček, J., and Mocek, T. Non-fluorinated superhydrophobic Al7075 aerospace alloy by ps laser processing. *Appl. Surf. Sci.*, 493, 2019, pp. 287–293, 10.1016/j.apsusc.2019.07.035

[56] Schlie, S., Fadeeva, E., Koroleva, A., Ovsianikov, A., Koch, J., Ngezahayo, A., and Chichkov, B. N. Laser-based nanoengineering of surface topographies for biomedical applications. *Photon. Nanostruct. - Fundam. Appl.*, 9, 2011, pp. 159–162, 10.1016/j.photonics.2010.09.006

[57] Jeong, Y.-H., Choe, H.-C., and Brantley, W. A. Nanostructured thin film formation on femtosecond laser-textured Ti–35Nb–xZr alloy for biomedical applications. *Thin Solid Films*, 519, 2011, pp. 4668–4675, 10.1016/j.tsf.2011.01.014

[58] Fasasi, A. Y., Mwenifumbo, S., Rahbar, N., Chen, J., Li, M., Beye, A. C., Arnold, C. B., and Soboyejo, W. O. Nano-second UV laser processed microgrooves on Ti6Al4V for biomedical applications. *Mater. Sci. Eng. C*, 29, 2009, pp. 5–13, 10.1016/j.msec.2008.05.002

[59] Sanguedolce, M., Rotella, G., Siciliani, V., Pelaccia, R., Orazi, L., and Filice, L. Surface characterization of ultra-short laser textured titanium for biomedical application. *Procedia CIRP*, 110, 2022, pp. 128–132, 10.1016/j.procir.2022.06.024

[60] Rajendran, A. and Pattanayak, D. K. Bioactive and antimicrobial macro-/micro-nanoporous selective laser melted Ti–6Al–4V alloy for biomedical applications. *Heliyon*, 8, 2022, 10.1016/j.heliyon.2022.e09122

[61] Suresh, S., Sun, C.-N., Tekumalla, S., Rosa, V., Nai, S. M. L., and Wong, R. C. W. Mechanical properties and in vitro cytocompatibility of dense and porous Ti–6Al–4V ELI manufactured by selective laser melting technology for biomedical applications. *J. Mech. Behav. Biomed. Mater.*, 123, 2021, p. 104712, 10.1016/j.jmbbm.2021.104712

[62] Hammadi, O. A. Using third-harmonic radiation of Nd: YAG laser to fabricate high-quality microchannels for biomedical applications. *Optik*, 208, 2020, p. 164147, 10.1016/j.ijleo.2019.164147

[63] Yan, X. J., Gugel, H., Huth, S., and Theisen, W. Microstructures and properties of laser cladding NiTi alloy with W for biomedical applications. *Mater. Lett.*, 65, 2011, pp. 2934–2936, 10.1016/j.matlet.2011.06.040

[64] Zhang, X., Pfeiffer, S., Rutkowski, P., Makowska, M., Kata, D., Yang, J., and Graule, T. Laser cladding of manganese oxide doped aluminum oxide granules on titanium alloy for biomedical applications. *Appl. Surf. Sci.*, 520, 2020, p. 146304, 10.1016/j.apsusc.2020.146304

[65] Ibrahim, M. Z., Sarhan, A. A. D., Kuo, T. Y., Yusof, F., and Hamdi, M. Characterization and hardness enhancement of amorphous Fe-based metallic glass laser cladded on nickel-free stainless steel for biomedical implant application. *Mater. Chem. Phys.*, 235, 2019, p. 121745, 10.1016/j.matchemphys.2019.121745

[66] Tsigkis, V., Rahman, M. S., Hackel, L., Davami, K., Beheshti, A., and Polycarpou, A. A. Helium tribology of Inconel 617 subjected to laser peening for high temperature nuclear reactor applications. *Appl. Surf. Sci.*, 577, 2022, p. 151961, 10.1016/j.apsusc.2021.151961

[67] Zhu, J., Jiao, X., Zhou, C., and Gao, H. Applications of underwater laser peening in nuclear power plant maintenance. *Energy Procedia*, 16, 2012, pp. 153–158, 10.1016/j.egypro.2012.01.026

[68] Baldridge, T., Poling, G., Foroozmehr, E., Kovacevic, R., Metz, T., Kadekar, V., and Gupta, M. C. Laser cladding of Inconel 690 on Inconel 600 superalloy for corrosion protection in nuclear applications. *Opt. Lasers Eng.*, 51, 2013, pp. 180–184, 10.1016/j.optlaseng.2012.08.006

[69] Pacquentin, W., Gouton, L., Caron, N., Brussieux, C., Foucault, M., Peyre, P., Maskrot, H., and Favier, V. Laser surface melting of nickel-based alloy reduces nickel release in the primary cooling system of a nuclear power plant. *Opt. Laser Technol.*, 144, 2021, p. 107401, 10.1016/j.optlastec.2021.107401

[70] Zhang, Y., Zhou, Y. J., Lin, J. P., Chen, G. L., and Liaw, P. K. Solid-solution phase formation rules for multi-component alloys. *Adv. Eng. Mater.*, 10, 2008, pp. 534–538, 10.1002/adem.20070024

[71] Sheng, G. U. O. and Liu, C. T. Phase stability in high entropy alloys: Formation of solid-solution phase or amorphous phase. *Prog. Nat. Sci. Mater. Int.*, 21, 2011, pp. 433–446, 10.1016/S1002-0071(12)60080-X

[72] Ren, M.-x., Li, B.-s., and Fu, H.-z. Formation condition of solid solution type high-entropy alloy. *Trans. Nonferr. Metal. Soc.*, 23, 2013, pp. 991–995, 10.1016/S1003-6326(13)62557-1

[73] Guo, S., Ng, C., Lu, J., and Liu, C. T., Effect of valence electron concentration on stability of FCC or BCC phase in high entropy alloys. *J. Appl. Phys.* 109, 2011, p. 103505, 10.1063/1.3587228

[74] Zhang, Y., Yang, X., and Liaw, P. K. Alloy design and properties optimization of highentropy alloys. *JOM*, 64, 2012, pp. 830–838, 10.1007/s11183 7-012-0366-5

Chapter 10

Laser Processing Technologies in Electronic and MEMS Packaging for Advancement of Industry 4.0

Payal Bansal, Kalpit Jain, and Chetan Dudhagara

CONTENTS

10.1 Introduction 235
    10.1.1 Role of Sensors 236
    10.1.2 Advancement 237
10.2 The Evolution of Biosensors 237
    10.2.1 Essential Idea of Biosensors 237
    10.2.2 Nano Materials 238
    10.2.3 AI Biosensors 238
    10.2.4 Flexible Bioelectronics Materials and Integration 239
    10.2.5 Wireless Communication 239
    10.2.6 Machine Learning 239
    10.2.7 Smartphone-Based Platform 239
    10.2.8 Designs in the Semiconductor Enterprises and
            Packaging Foundries 240
10.3 Laser Drilling 242
    10.3.1 Laser Cutting 243
    10.3.2 Accentuation 243
10.4 Conclusion and Future Scope 244
10.5 Future Scope 245
References 245

10.1 INTRODUCTION

Inferable from the clever region at the association point among the straightforward central correspondence structures, sensors go about as perfect statistics getting maneuver for the affirmation of adroit metropolitan networks. Considering use circumstances, sensors can be requested into genuine sensors, engineered, and biosensors. As one of the key portrayals in overall structure, biosensors have encountered a long improvement from superb electrochemical biosensors to wearable and implantable biosensors and have been for the most part applied in food security, clinical consideration, disorder assurance, normal checking, and biosafety. The fundamental engineering of computer-based intelligence biosensors is made out of three primary components: data assortment, signal transformation, and artificial intelligence (AI) information handling [1–5].

In laser machining, the laser is utilized as an energy hotspot for material evacuation. The material expulsion can happen by means of a warm course, for example, dissolving and vanishing, or a non-warm course, as in direct substance bond breaking by separation and decay, which at last prompts material discharge from the objective surface. Laser machining is worthwhile contrasted and customary mechanical machining as the hardness of the material needn't bother with to be thought of. The powers engaged with laser machining are tiny as the strain applied by the photons is immaterial for mass material. Then again, the optical and warm properties like absorptivity, reflectivity, and warm diffusivity should be considered to decide how much material is eliminated. On top of this, the laser handling boundaries should likewise be examined to acquire the ideal machining result [6–9].

CO_2 and Nd:YAG lasers are the most usually involved gear for laser handling applications. The qualifications among the bar assets and nature of the two laser types and the different material affiliations unequivocally conclude their relevance fields. The frequency of the shaft created by a CO_2 laser is about IO pn, which is ingested well by plastics, wood, paper, and organics. The glass is obscure at this frequency too. Aside from the way that metals almost behave like mirrors and retain a couple of percent of the medium-IR shaft, sheet-metal cutting and welding is the biggest market for CO_1 laser handling frameworks: around 20,000 frameworks have been introduced for this application around the world. This is thanks to the high constant (normal) yield force of CO_2 lasers. The below force of Nd:YAG lasers working in beat mode gives low cutting and welding speed, yet it actually enjoys a huge benefit: their high energy heartbeats can defeat the surface reflectance of most materials, in this way giving compelling penetrating, etching, and different cycles [10–12].

10.1.1 Role of Sensors

There are two important areas where individual patients get the benefit of suitable healthcare as AI and wearable sensors. The combination of these areas makes it possible for somebody to improve the procurement of patient information and further create a plan of wearable sensors for noticing the wearer's prosperity, well-being, and ecological elements. Right now, as the web of things (IoT), large information and enormous well-being move from idea to execution, man-made intelligence biosensors with proper specialized characteristics are confronting new open doors and difficulties. In the research work, the furthermost developed advancement through the vital stages for future innovation from biosensing, wearable biosensing, to simulate intelligence biosensing is summed up. Indeed, material advancement, biorecognition component, signal securing and transportation, information handling, and insight choice framework are the main parts, which are the principal focal point of conversation [13–15]. The difficulties and chances of AI biosensors pushing ahead toward future medication gadgets are additionally talked about.

10.1.2 Advancement

As of late, biosensing has entered another stage because of the advancement and execution of ideas, for example, enormous health, IoT, and big data. The creation of wearable biosensors intends to get over the impediments of concentrated, responsive medical care by giving people an understanding of their own actual dynamic. Be that as it may, complex causality brings about outrageous trouble in the result examination. Joining of man-made brainpower (AI) approaches including design examination and characterization calculations with biosensors can overcome any issues between the information procurement and investigation and accomplish work on demonstrative and restorative exactness. This survey is expected to present the latest advancement in computer-based intelligence biosensors and talked about the materials, remote correspondence, AI, and choice applied in simulated intelligence biosensors in the beyond quite a while. We likewise featured the conceivable future man-made intelligence biosensors and related progressed programming capability including man-made intelligence conclusion, enormous information handling, and self-learning/adaption [16–18]. Such man-made intelligence biosensors include cross-breed strategies of remote biosensing innovation and high-level AI calculations, holding extraordinary commitment for acknowledging consistent checking of medical services and cloud-associated point of care diagnostics.

10.2 THE EVOLUTION OF BIOSENSORS

10.2.1 Essential Idea of Biosensors

The biosensors are a device that is fragile to natural constituents and transforms the attentiveness into signals for ID. As demonstrated by the Worldwide Relationship of Pure and Applied Science, a biosensor is an integrated receptor-transducer device to give specific quantitative or semi-quantitative wise information about a specific examine, which is schematically contained biosensitive materials as a bio acknowledgment part, a physical or substance transducer part, a sign transmission, and enhancer part. As per the standard perspective, considering the sort of transducer units, biosensors can be described as bio cathode sensors, semiconductor biosensors, warm bio-sensors, photo biosensors, and piezoelectric valuable stone biosensors. Considering the sort of conspicuous verification parts, biosensors can be disengaged into impetus sensors, nucleic destructive sensors, and safe sensors. Considering the kind of affirmation parts, biosensors can be gathered into bio partiality biosensors, metabotropic biosensors, and reactant biosensors. The chance of protein cathodes was first proposed by Clark and Lyons considering the compound Gox over an aerometric oxygen terminal, which set out on the important splendid season of biosensors. Different biomarkers and biomaterials were picked as affirmation parts of biosensors, including compounds, antibodies, nucleic acids, cells, tissue sheets, and microorganisms. New biosensor transducers were similarly introduced during a decade, including electro-chemical biosensors, warm biosensors, semiconductor biosensors, optical fiber biosensors, piezoelectric, mass, and acoustic biosensors. The second improvement of biosensors was gigantic degree promoted applications. The Yellow Spring Instruments (Ohio) glucose analyzers watched out for two critical progressions for the reputation of early impetus terminals. The principal headway was to use polycarbonate film development to diminish the grouping of glucose; the other was the immobilization of GOx by cross-interfacing. The affirmation of nucleic destruction pro-voked the improvement of DNA biosensors and DNA microarrays in the last piece of the 1990s [19–24].

10.2.2 Nano Materials

Nano materials have alluring assets; at the smallest to smallest size, the responsiveness of the sensor works as the level of particles on the outer layer of the sensor edge fundamentally increments. Smallest dimension of the particle impact can prompt changes in the properties, particularly rea-sonable for the investigation of natural cycles in living cells and the improvement of significant illnesses.

10.2.3 AI Biosensors

Over the course of the last ten years, there has been a developing accentuation on implanting shrewdness on the planet. The canny handling of biosensing data has transformed biosensors. The mix of computer-based intelligence and bio-sensors has constructed an idea of computer-based intelligence biosensors. The fundamental engineering of computer-based intelligence biosensors is made out of three primary components: data assortment, signal transformation, and AI information handling. Data assortment alludes to a gathering of bio-sensors for constant checking of physical, synthetic, natural, climate, or personality data.

10.2.4 Flexible Bioelectronics Materials and Integration

Adaptable bioelectronic resources, for example, adaptable movies, materials, swathes, spots, and tattoos assist as machine-driven services for coordinating electronic circuits which assume a significant part in AI biosensors. Polyimide and polyethylene terephthalate are recorded broadly utilized as adaptable bioelectronics materials in biosensors coordination and innovation, attributable to their great exhibition, client solace, high oxygen penetrability, and similarity with roll-to-move manufacture processes.

10.2.5 Wireless Communication

The AI biosensors are preferably connected with remote information correspondence to communicate the data among biosensors and cell phone-based stages or other smart terminals. Simulated intelligence biosensors in light of remote innovation without link can lessen the expense and continually add new gadgets and adaptability. Remote innovations zeroed in on artificial intelligence biosensor network (AIBN) with mass market reception are RFID, NFC, Wi-Fi, and ZigBee.

10.2.6 Machine Learning

AI biosensors are important just when information can be utilized by us, and understanding what the information implies is a prerequisite for utilizing them. AI information handling includes gaining from the information, including how to break down information around, reach the right determinations from the information, and perceive in the event that the information has been confused [25–27].

10.2.7 Smartphone-Based Platform

Cell phone-based detecting frameworks stand out with the overall prominence of cell phones. Because of the coordination of various sensors and works like handling and correspondence, cell phone-based stages are assuming a significant part in man-made intelligence biosensors for information handling, sharing, stockpiling, and communication with cloud. With a surge of cell phone applications amassing into the area of simulated intelligence biosensors for handling information returned by add-on biosensing modules, the utilization of cell phone-based stages is at this point not a clever subject. These cell phone-based detecting frameworks frequently comprise a cell phone and extra biosensing modules gadgets like electrochemistry, fluorescence, and plasmon reverberation or haul around an additional equipment part, for example, cameras, Bluetooth, USB, and

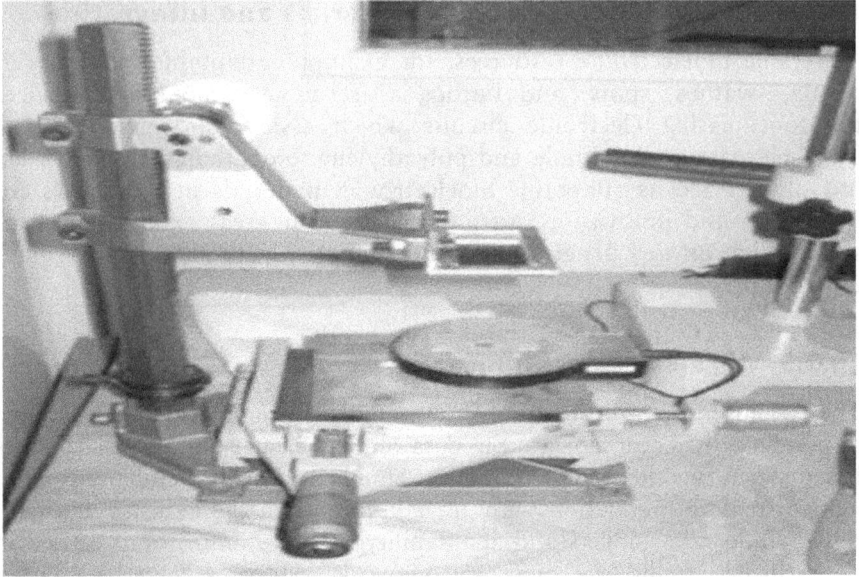

Figure 10.1 Manual stencil printer stage with sample holder.

sound port, to increment openness, control the identification cycle, and get the discovery information (see Figure 10.1).

Electronic bundling advances continue changing in view of the continuous interest. The microelectronics business is ceaselessly searching for higher thickness bundling and more chip usefulness. This outcomes in greater chips yet more modest bundles, so the bite of the dust is turning into a bigger piece of the all out bundle volume. Figure 10.2 shows the progression of the main level bundles in the latest years [28,29]. The steadily expanding interest in more chip highlights in more modest bundles drives future bundling innovation. All the more as of late, three-layered (3D) stacked bite the dust bundling system is embraced to moderate a portion of these difficulties. All the downsizing techniques for electronic bundling displayed in Figure 10.3 require productive manufacturability and cost that end up being even more troublesome. Two normally used scaling-down processes are integral metal–oxide–semiconductor (CMOS) scaling and bundling scaling. Be that as it may, the bundling scene is changing rapidly toward framework-level scaling because of greater expense and assembling difficulties of CMOS scaling. Figure 10.4 exhibits the change in outlook for framework-level scaling.

10.2.8 Designs in the Semiconductor Enterprises and Packaging Foundries

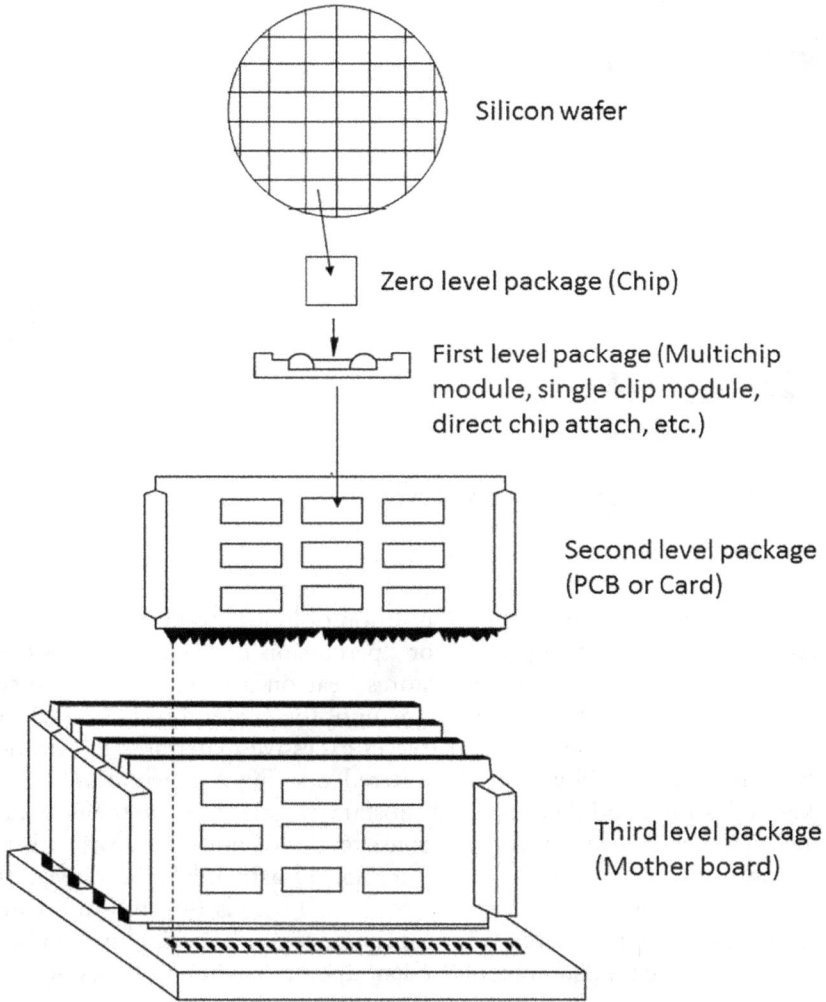

Silicon wafer

Zero level package (Chip)

First level package (Multichip module, single clip module, direct chip attach, etc.)

Second level package (PCB or Card)

Third level package (Mother board)

Figure 10.2 Hierarchy of electronic packaging.

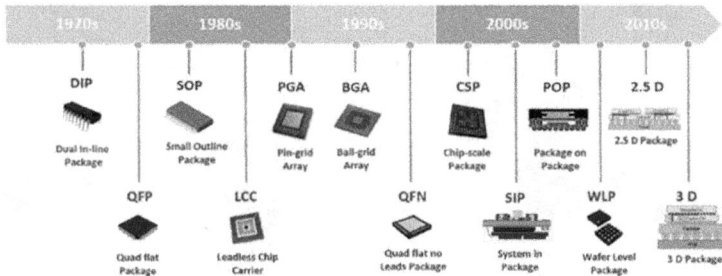

Figure 10.3 Development of the first-level packages in the last decades.

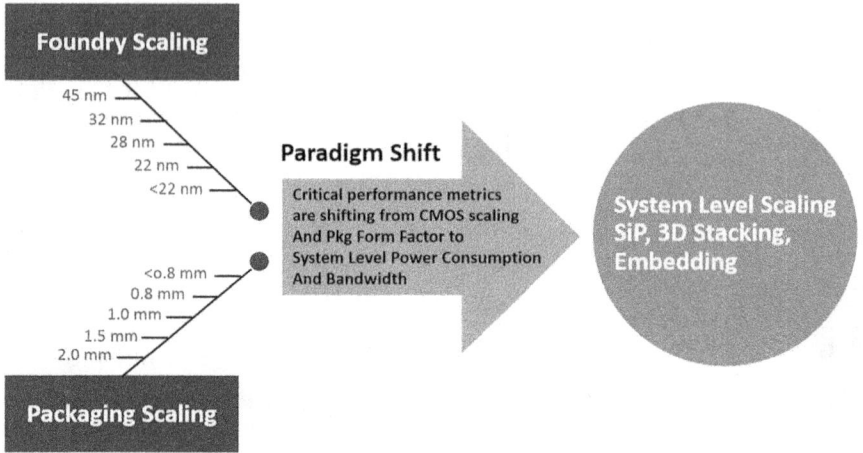

Figure 10.4 IC packaging—the changing landscape.

10.3 LASER DRILLING

Laser drilling system is used for the creation of holes through the laser, recommended as "popped" openings or "percussion depleted" openings, by utilizing a centered beat laser that stores heat on a material, hence disintegrating level to level (see Figure 10.5) until the creation of holes [10,11]. High amount of heat plays a vital role in excessive material. Vaporization makes the material volume in the entered opening augmentation, which makes high strain, and then the fluid substantial start to create the holes. All around, laser shaft width at focus is consistently 0.2 mil (5 lm), which is the estimation of the "laser exhausting mechanical assembly". Moreover, the typical required opening distances across are 1.6 mils (40 lm) and 4 mils (100 lm). The opening between these openings can be basically just 2 mils, or impressively more unassuming, dependent upon the material and cycle used.

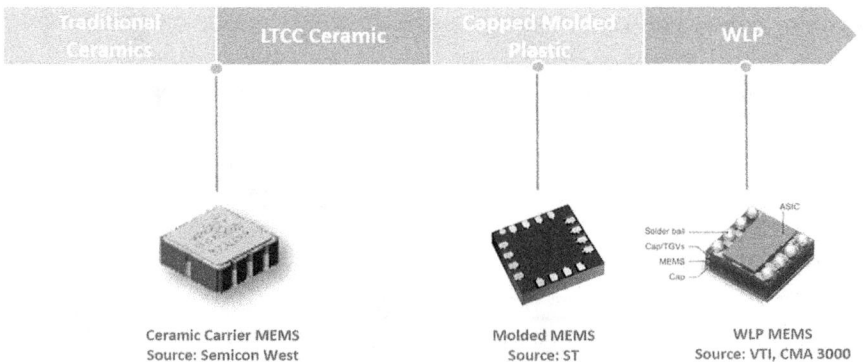

Figure 10.5 MEMS packaging—the changing landscape.

Single-shot Drilling   Percussion Drilling   Trepanning Drilling   Helical Drilling

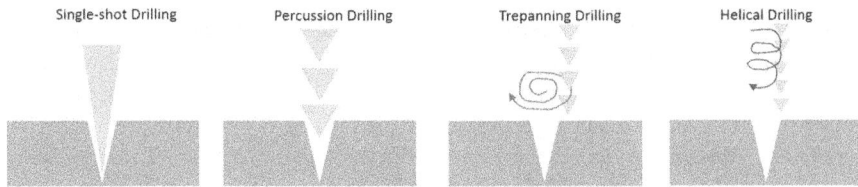

Figure 10.6 A schematic of several laser drilling processes.

In the occasion that more noteworthy creation of holes is required, the place where the laser will directly contact the material moved with some restriction in a system known as trepanning, ideal broadness is made. Over the long run, several exhausting strategies [10] are brainstormed from the fundamental approach as shown by a schematic diagram in Figure 10.6. The laser drilling is very better as compared with mechanical, e.g., mechanical entering are punching, proposing, and wire electrical delivery machining, for instance, no drill breakage or contraption uniform due to the noncontact make, huge opening dimensions and profile because of the software design adaptableness, drill openings on determined and twisted surfaces due to the noncontact handle and multi pivot limit, noncontact cutting strategy have the ability to drill hard and delicate materials easily, charge vicious considering the rapid infiltrating cycles, no usable expenditures, and insignificant individual time.

10.3.1 Laser Cutting

The laser can play out different cutting endeavors going into the form of scale as it cuts the thin chip of semiconductor in the form of good quality chips [12–14]. In laser cutting, the laser of the machine directly impacts the materials and then it increases the temperature of materials, such a great deal of that it smooth or even deteriorates. Exactly when the work piece ingests a satisfactory proportion of heat, the machine will be ready to cut the materials. Normally, a surge of gas setbacks the required material of fluid in downward condition for cutting the material. The thickness of the cut is ordinarily negligible greater than the laser point of support estimation itself. Generally speaking, laser point of support width at emphasis is customarily 0.2 mil (5 lm), which is the estimation of the laser bore.

Bundle explicit cutting cycles, for example, laser extract and laser dicing are portrayed in Section 3.

10.3.2 Accentuation

Accentuation is located in the associated fields of versatile bioelectronic resources and compromise, remote and correspondence, computer-based intelligence, and cell-based stages to improve man-made reasoning biosensors [29–32]. Slimness, downsizing, compromise, and less excited for power are the inevitable destiny of man-made knowledge biosensors for clinical benefits. The maximum development in the experimental attempts ought to be concentrated on the acknowledgment of low-level, unequivocal biomarkers in bioliquids and structure concession. Biomarkers ought to defeat any hindrance between bio real markers and biochemical markers to get a complete structure. Many undertakings ought to be given to making versatile electronic resources for coordinating chip advancement, IoT, and computerized reasoning to recognize self-learning, self-get-together, and self-change of man-made knowledge biosensor structures. The gathered information will be taken care into AI calculations to screen fundamental signs, spot anomalies, and track medicines. What's more, taking into account the connection between human–machine regular interactions is fundamental. By creating AIBN, we cannot just take shut circle choices relying upon what they detect or the got input, yet additionally gain from the environment, past communication, and adaption. In spite of developing growth ending in the beyond couple of time in the area of simulated intelligence biosensors in the direction of medical sector, important difficulties stay formerly the business development of simulated intelligence biosensors [33,34]. First and foremost, adaptable bioelectronic materials are a basic variable for business solicitations. Hominoid groups and their inward creatures are characteristically adaptable and flexible. For this situation, it is vital for the reconciliation of gadgets in adaptable material stages. Inferable from very adaptable mechanical properties, adaptable bioelectronics carry benefits to coordinate the human body and organs with negligible mechanical harms to muscles and lessen incidental effects after long-haul incorporation. Additionally, the superb places of collaboration between versatile bioelectronics and structures that have, in proper organized form, empowered hail transfer with high sign to upheaval extent. Computer-based intelligence biosensors incorporated into a framework with restricted space are vital. We presume this survey can act as a standard to tackle the difficulties that at present occur for simulated intelligence biosensors. Via consistent framework, man-made intelligence biosensors will give another stage to future development and can possibly transform practically every part of medical care [35–38].

10.4 CONCLUSION AND FUTURE SCOPE

The packaging of electronic and microelectromechanical systems (MEMS) devices is an important part of the overall manufacturing process as it ensures mechanical robustness as well as required electrical/electromechanical functionalities. The packaging integration process involves the selection of packaging materials and technology, process design, fabrication, and testing. As the demand for functionalities of an electronic or MEMS device is increasing every passing year, chip size is getting larger and is occupying the majority of space within a package. This requires innovative packaging technologies so that integration can be done with less thermal/mechanical effect on the nearby components. Laser processing technologies for electronic and MEMS packaging have the potential to obviate some of the difficulties associated with traditional packaging technologies and can become an attractive alternative for the small-scale integration of components.

10.5 FUTURE SCOPE

- It will become smaller and less powerful.
- In some cases down to single photon emission.
- Lasers will get ever more powerful, up to a region where it pulls particles out of the vacuum.

REFERENCES

[1] Osseiran, A.; Boccardi, F.; Braun, V.; Kusume, K.; Marsch, P.; Maternia, M.; Queseth, O.; Schellmann, M.; Schotten, H.; Taoka, H.; et al. Scenarios for 5G Mobile and Wireless Communications: The Vision of the METIS Project. *IEEE Commun. Mag.* 2014, 52, 26–35. [CrossRef]

[2] Hwang, I. J.; Oh, J. I.; Jo, H. W.; Kim, K. S.; Yu, J. W.; Lee, D. J. 28 GHz and 38 GHz Dual-Band Vertically Stacked Dipole Antennas on Flexible Liquid Crystal Polymer Substrates for Millimeter-Wave 5G Cellular Handsets. *IEEE Trans. Antennas Propag.* 2022, 70, 3223–3236. [CrossRef]

[3] Hong, W. Solving the 5G Mobile Antenna Puzzle: Assessing Future Directions for the 5G Mobile Antenna Paradigm Shift. *IEEE Microw. Mag.* 2017, 18, 86–102. [CrossRef]

[4] Gowrish, B.; John, D.; Settu, D.; Basu, A.; Koul, S. K. Novel Mechanical Reconfigurable PCB Antenna for 2.4 GHz Wireless Consumer Product: Minimizing Time to Market. In Proceedings of the 2014 3rd Asia-Pacific Conference on Antennas and Propagation, Harbin, China, 26–29 July 2014; pp. 18–21.

[5] Lee, G. H.; Moon, H.; Kim, H.; Lee, G. H.; Kwon, W.; Yoo, S.; Myung, D.; Yun, S. H.; Bao, Z.; Hahn, S. K. Multifunctional Materials for Implantable and Wearable Photonic Healthcare Devices. *Nat. Rev. Mater.* 2020, 5, 149–165. [CrossRef]

[6] Zhang, H.; Lan, Y.; Qiu, S.; Min, S.; Jang, H.; Park, J.; Gong, S.; Ma, Z. Flexible and Stretchable Microwave Electronics: Past, Present, and Future Perspective. *Adv. Mater. Technol.* 2020, 32, 2000759. [CrossRef]

[7] Dudala, S.; Dubey, S. K.; Goel, S. Fully Integrated, Automated, and Smartphone Enabled Point-of-Source Portable Platform with Microfluidic Device for Nitrite Detection. *IEEE Trans. Biomed. Circuits Syst.* 2019, 13, 1518–1524. [CrossRef]

[8] Xiong, W. N.; Zhu, C.; Guo, D. L.; Hou, C.; Yang, Z. X.; Xu, Z. Y.; Qiu, L.; Yang, H.; Li, K.; Huang, Y. A. Bio-Inspired, Intelligent Flexible Sensing Skin for Multifunctional Flying Perception. *Nano Energy* 2021, 90, 106550. [CrossRef]

[9] Chen, K.; Yao, Y. Process Optimisation in Pulsed Laser Micromachining with Applications in Medical Device Manufacturing, *Int. J. Adv. Manuf. Technol.* 2000, 16, 243–249.

[10] Pham, D.; Dimov, S.; Petkov, P. Laser Milling of Ceramic Components, *Int. J. Mach. Tool Manuf.* 2007, 47, 618–626.

[11] Tatah, A.; Fukumoto, A. 3-Dimensional Micromachining with Femtosecond Laser Pulses, U.S. Patent 5,786,560, 28 7, 1998.

[12] Sun, S.; Brandt, M.; Dargusch, M. Thermally Enhanced Machining of Hard-to-Machine Materials—A Review, *Int. J. Mach. Tool Manuf.* 2010, 50, 663–680.

[13] Rebro, P.; Shin, Y. C.; Incropera, F. P. Laser-Assisted Machining of Reaction Sintered Mullite Ceramics, *J. Manuf. Sci. Eng.* 2002, 124, 875–885.

[14] Lei, S.; Shin, Y. C.; Incropera, F. P. Experimental Investigation of Thermo-Mechanical Characteristics in Laser-Assisted Machining of Silicon Nitride Ceramics, *J. Manuf. Sci. Eng.* 2001, 123, 639–646.

[15] Rozzi, J. C.; Pfefferkorn, F. E.; Incropera, F. P.; Shin, Y. C. Transient, Three-Dimensional Heat Transfer Model for the Laser Assisted Machining of Silicon Nitride. I. Comparison of Predictions with Measured Surface Temperature Histories, *Int. J. Heat Mass Transf.* 2000, 43(8), 1409–1424.

[16] Pfefferkorn, F. E.; Shin, Y. C.; Tian, Y.; Incropera, F. P. Laser-Assisted Machining of Magnesia-Partially Stabilized Zirconia, *J. Manuf. Sci. Eng.* 2004, 126, 42–51.

[17] Pajak, P.; Desilva, A.; Harrison, D.; Mcgeough, J. Precision and Efficiency of Laser Assisted Jet Electrochemical Machining. *Precis. Eng.* 2006, 30, 288–298.

[18] Rajurkar, K.; Kozak, J. Laser Assisted Electrochemical Machining, *Trans. NAMRI* 2001, 29, 421–427.

[19] DeSilva, A.; Pajak, P.; Harrison, D.; McGeough, J. Modelling and Experimental Investigation of Laser Assisted Jet Electrochemical Machining, *CIRP Ann.* 2004, 53(1), 179–182.

[20] Sankar, M.; Gnanavelbabu, A.; Rajkumar, K. Effect of Reinforcement Particles on the Abrasive Assisted Electrochemical Machining of Aluminium-Boron Carbide-Graphite Composite. *Procedia Eng.* 2014, 97, 381–389.

[21] Hnatovsky, C.; Taylor, R. S.; Simova, E.; Bhardwaj, V. R.; Rayner, D. M.; Corkum, P. B. Polarization-Selective Etching in Femtosecond Laser-Assisted Microfluidic Channel Fabrication in Fused Silica. *Opt. Lett.* 2005, 30(14), 1867–1869.

[22] Maselli, V.; Osellame, R.; Cerullo, G.; Ramponi, R.; Laporta, P.; Magagnin, L.; Cavallotti, P. L. Fabrication of Long Microchannels with Circular Cross Section Using a Stigmatically Shaped Femtosecond Laser Pulses and Chemical Etching. *Appl. Phys. Lett.* 2006, 88, 191107.

[23] Sugioka, K.; Akane, T.; Obata, K.; Toyoda, K.; Midorikawa, K. Multi Wavelength Excitation Processing Using F2 and KrF Excimer Lasers for Precision Microfabrication of Hard Materials. *Appl. Surf. Sci.* 2002, 197, 814–821.

[24] Obata, K.; Sugioka, K.; Akane, T.; Midorikawa, K.; Aoki, N.; Toyoda, K. Efficient Refractive-Index Modification of Fused Silica by a Resonance-Photoionization-Like Process Using F2 and KrF Excimer Lasers. *Opt. Lett.* 2002, 27(5), 330–332.

[25] Sugioka, K.; Meunier, M.; Pique, A. *Laser Precision Microfabrication*. Springer-Verlag: Berlin, Heidelberg, 2010.

[26] Rubenchik, A.M.; Feit, M. D. Initiation, Growth, and Mitigation of UV-Laser-Induced Damage in Fused Silica, *Proc. SPIE* Laser-Induced Damage in Optical Materials: 2001, 2002, 4679. 10.1117/12.461680.

[27] Bloembergen, N. Laser Induced Electric Breakdown in Solids, *IEEE J. Quantum Electron.* 1974, 10(3), 375–386.

[28] Genin, F.; Salleo, A.; Pistor, T.; Chase, L. Role of Light Intensification by Cracks in Optical Breakdown on Surfaces, *J. Opt. Soc. Am. A* 2001, 18(10), 2607–2616.

[29] Patil, P.; Kumar, A.; Gori, Y. FEA Design and Analysis of Flow Sensor. *Mater. Today: Proc.* 2021, 46, Part 20, 10563–10568, ISSN 2214-7853, 10.1016/j.matpr.2021.01.315

[30] Patil, P.; Kumar, A.; Gori, Y. Coriolis Mass Flow Sensor (CMFS): A Review. *J. Critic. Rev.* 2019, 6(6), 2628–2632, ISSN-2394-5125.

[31] Patil, P. P.; Kumar, A. Finite Element Analysis of Omega Type Coriolis Mass Flow Sensor (CMFS) for Evaluation of Fundamental Frequency and Mode Shape. *Int. J. Eng. Technol.* 2017, 9(3). 10.21817/ijet/2017/v9i3/170903S017

[32] Patil, P. P.; Kumar, A. Design and FEA Simulation of Omega Type Coriolis Mass Flow Sensor. *Int. J. Control Theory Appl.* 2017, 9(40), 383–387.

[33] Patil, P. P.; Kumar, A.; Ahmad, F. Omega Design and FEA Based Coriolis Mass Flow Sensor (CMFS) Analysis Using Titanium Material. *IOP Conf. Series: Mater. Sci. Eng.* 2017, 310(2018), 012005. doi:10.1088/1757-899X/310/1/012005.

[34] Patil, P.; Sharma, S.; Saini, A.; Kumar, A. ANN Modelling of Cu Type Omega Vibration Based Mass Flow Sensor. *Procedia Technol.* 2014, 14, 260–265. DOI: 10.1016/j.protcy.2014.08.034

[35] Kumar, A.; Datta, A.; Kumar, A. Recent Advancements and Future Trends in Next-Generation Materials for Biomedical Applications. In *Advanced Materials for Biomedical Applications*; Kumar, A.; Gori, Y.; Kumar, A.; Meena, C. S.; Dutt, N., Eds.; CRC Press: Boca Raton, FL, USA, 2022; Chapter 1; pp. 1–19.

[36] Prasad, A.; Chakraborty, G.; Kumar, A. Bio-Based Environmentally Benign Polymeric Resorbable Materials for Orthopedic Fixation Applications. In *Advanced Materials for Biomedical Applications*; Kumar, A.; Gori, Y.; Kumar, A.; Meena, C. S.; Dutt, N., Eds.; CRC Press: Boca Raton, FL, USA, 2022; Chapter 15; pp. 251–266.

[37] Datta, A.; Kumar, A.; Kumar, A.; Kumar, A.; Singh, V. P. Advanced Materials in Biological Implants and Surgical Tools. In *Advanced Materials for Biomedical Applications*; Kumar, A., Gori, Y.; Kumar, A.; Meena, C. S.; Dutt, N., Eds.; CRC Press: Boca Raton, FL, USA, 2022; Chapter 2; pp. 21–43.

[38] Kumar, A.; Gangwar, A. K. S.; Kumar, A.; Meena, C. S.; Singh, V. P.; Dutt, N.; Prasad, A.; Gori, Y. Biomedical Study of Femur Bone Fracture and Healing. In *Advanced Materials for Biomedical Applications*; Kumar, A.; Gori, Y.; Kumar, A.; Meena, C. S.; Dutt, N., Eds.; CRC Press: Boca Raton, FL, USA, 2022; Chapter 14; pp. 235–250.

Index

3D Printing, 23, 26, 57, 60, 63, 85
4D Printing, 58, 62

Accentuation, 243
Additive Manufacturing (AM), 20, 55, 56, 62, 68, 87, 91, 108, 144, 165
Aerospace, 60, 78, 159, 219
AI Biosensors, 238
Amplification, 50
Angioplasty, 192
ANOVA, 99
Artificial Intelligence (AI), 85, 86, 102, 194, 198, 236
ASTM, 57, 159
Atmospheric Plasma Sprayed (APS), 223
Automation, 56
Automobile, 77, 218

Batteries, 87
Beam, 25
Bioelectronics, 239, 244
Bio-Implants, 188
Biomedical, 176, 195, 220
Biomaterials, 59
Biosafety, 236
Bio-Sensors, 191, 237, 238
Building, 60

CAD, 56, 57
Carbon Dioxide (CO_2) Laser, 20, 182, 236
CCD (Central Composite Design), 161
Ceramics, 178
Challenges, 80
Clinical, 34
Common Industrial Lasers, 181
Components, 61, 73

Computer-Aided Manufacturing (CAM), 120
Concave, 10, 12
Confocal Scanning Microscopy, 33
Control, 59
Convex, 10, 11Convolutional Neural Network (CNN), 88, 90, 92, 100
Cutting, 107
CNC, 81

Data acquisition, 59
Deep learning (DL), 85
Defects, 29, 93, 75, 125
Defense, 109
Dental Applications, 77
Diffuse Reflector, 07
Diode, 17
Direct Digital Manufacturing (DDM), 56
Directed Energy Deposition (DED), 94
DNA biosensors, 238
Doped, 22
Drill Bit, 190
Drilling, 112

Efficiency, 61, 63
Electrochemical, 235
Electron Beam Melting (EBM), 74
Electronics, 235
Electronics levels, 46
Energy level, 46
Energy efficient, 111
Evaporation, 136
Excitation, 48
Experimental design, 30

Fiber, 22
Fiber laser machining, 92

Fixtures, 117
Flow cytometry, 34
Fluorescence correlation, 33
Fused deposition modeling (FDM), 60

Gas-discharge lasers, 14
Gas laser, 23

Heat-affected zone (HAZ), 75, 176
High Entropy Alloy (HEA), 207, 209,
        213, 220, 221, 222, 223, 225,
        226, 227, 228
Hot isostatic pressing (HIP), 74
Hybrid process, 58
Hybridization, 162

IEEE, 86
Industrial, 31, 81
Industry 4.0, 235
IoT, 85, 236, 244

Laminated Object Manufacturing
        (LOM), 69
Laser, 03, 13, 17, 18, 20, 22, 23, 52, 68,
        107, 207, 228
Laser Additive Manufacturing
        (LAM), 158
Laser-Arc Hybrid Additive
        Manufacturing
        (LAHAM), 163
Laser cladding, 211
Laser cleaning, 134
Laser cutting, 110, 117, 243
Laser-Directed Energy Deposition, 160
Laser drilling, 143, 242
Laser Engineered Net Shaping
        (LENS), 126
Laser glazing, 216
Laser-Guided Net Engineering
        (LENS), 28
Laser hardening, 213
Laser machining, 90
Laser–material interaction (LMI), 216
Laser metal deposition, 19
Laser micromachining (LMM), 170,
        173, 177, 199
Laser peening, 74
Laser Powder Bed Fusion, 123, 159
Laser sintering, 71
Laser Shock Peening (LSP), 208, 212
Laser surface alloying, 215
Laser surface melting (LSM), 208, 209,
        213, 220, 224, 226, 227, 228

Laser surface texturing (LST), 208
Laser threshold, 52
Laser welding, 118
Layer-by-layer, 57, 67
Layered manufacturing, 130
Limitations, 62

Machine industry, 111
Machine learning (ML), 88, 90, 97,
        102, 161, 239
Macro-scale, 29
Magnetic Abrasive Finishing (MAF), 75
Marking, 31
Materials, 57
Measurement, 32
Medical applications, 34, 77, 109
Medical devices, 137
Medicines, 185
Metal-based, 70
Micro electromechanical systems
        (MEMS), 170, 235, 245
Micro manufacturing, 58
Microarray scanning, 33
Microprocessor, 170
Microstructural, 125
Military, 78
Mirror and lenses, 09
Mitigation of stresses, 163
Monitoring, 58

Nanomaterials, 238
Neural Network, 89, 91
Noncontact, 32
Nondestructive, 91
Nuclear, 220

Optics, 4
Optics geometry, 5
Optimization, 58, 160, 226
Orthopedics, 61
Oscillation, 50

Pacemaker, 139
Packaging, 235, 240
Parabola, 11
PCB, 241
Photolithography, 31
Photons, 157
Piercing, 112
Plasma, 213, 223, 224
Polymer, 69
Population Inversions, 49
Porosity, 125

Post-processing, 73
Powder bed fusion, 68
Powder contamination, 30
Power Pulse Energy Intensity, 24
Precision, 110
Pre-processing software, 30, 85
Process, 56
Prosthetics, 192
Protective structures, 60
Prototype, 117
Pulse duration, 25
Pulsed thermography (PT), 92

Rapid prototyping (RP), 56
Ray, 06
Reflection, 06
Refraction, 08
Remanufacturing, 61
Robotics, 96
Rotational Levels, 48

Scabbling, 132
Scribing, 31
Selective Laser Melting (SLM), 28
Selective Laser Sintering (SLS), 27
Semiconductor, 17, 240
Sensors, 235
Sheet lamination, 129
Silicon, 177
Single pulse drilling, 114
Smartphone, 239
Solid-state, 21
Solid state lasers, 184
Solution heat treatment (SHT), 74
Species, 45
Spectroscopy, 33
Spontaneous emission, 48

Spot Size, 25
Standard, 57
Standardization, 57
Stents, 190
Stereolithography (SLA), 26, 56, 69
Structural design, 165
Substrate, 225
Surgical applications, 34
Sustainability, 59
Surgical tools, 189
Surface modification, 209
Surface treatment, 108, 145

Taguchi, 161
Technique, 57
Telecommunication, 109
Texturing, 132, 173
Tissue, 238
Tool Wear, 111
Transitions, 46
Two-photon polymerisation, 130

Ultrasonic, 70

Vaporization, 198
VAT Photopolymerization, 128
Vibrational Levels, 46

Wavelength, 24
Wearable, 235
Wire Stripping, 139
Wireless Communication, 239

X-ray computed tomography (XCT), 91

YAG Laser, 21, 236

For Product Safety Concerns and Information please contact our EU
representative GPSR@taylorandfrancis.com
Taylor & Francis Verlag GmbH, Kaufingerstraße 24, 80331 München, Germany